Fundamentals of Fusion Welding Technology

IFS

ADVANCED WELDING SYSTEMS

1

FUNDAMENTALS OF FUSION WELDING TECHNOLOGY

Jean Cornu
(English Language Editor: John Weston

Springer-Verlag Berlin Heidelberg GmbH

British Library Cataloguing in Publication Data

Advanced welding systems.
  Vol. 1: Fundamentals of fusion welding technology.
  1. Welding
  I. Cornu, Jean   II. Weston, John
  III. Soudage par fusion en continu, Vol. 1 Notions Fondamentales, *English*
  671.5'2   TS277

ISBN 978-3-662-11051-5   ISBN 978-3-662-11049-2 (eBook)
DOI 10.1007/978-3-662-11049-2

This book was first published in French in 1985 by Hermes, 51 rue Rennequin, 75017 Paris, under the title: 'Soudage par fusion en continu, Vol. 1 Notions Fondamentales'.

Translated by Sue Greener. English language edited by John Weston.
Phototypeset by Systemset, Stotfold, Bedfordshire.

FOREWORD

THE ORIGINS of welding are buried in the depths of antiquity, commencing with the forging of native gold and copper, progressing in the bronze age with the braze welding of castings, but not developing greatly until relatively recently. It has been this century, and the latter half in particular, that welding has developed to the stage where there are more than 100 variants. Furthermore, joining by welding has become such an efficient technique that much of our modern way of life would not be possible without it. The giant oil rigs, built to withstand the rigours of the North Sea, the minute wire connectors in the computer and transistor, and the automobile and truck, could not exist were it not for welding processes.

Originally a uniquely manual process, the needs of industry have this century required welding techniques which could be mechanised. Some processes, such as friction welding, were readily mechanised but the most flexible and adaptable fusion processes awaited developments which allowed a continuous wire to be rapidly fed into the fusion zone. These processes, such as MAG and submerged arc, rapidly gave rise to machines for welding, with many appearing before the Second World War.

However, it has been the advent of the industrial robot, a machine which can manipulate the welding process with some of the flexibility of man, which has catalysed and accelerated the rate of automating welding. The robot became possible because of developments in electronics, microwelding and computing, techniques which have also allowed developments in the technique, the machinery and the control of the fusion welding process itself.

The introduction of these machines and processes into factory environments present a number of problems. Typically, the choice of process, the type of machine, and the method of integration into production. The answers to such problems require an understanding of both the processes and the machinery. The French experts and authors of this series, J. Cornu, J. M. Detriche and P. Marchal, have considered the fusion welding processes and welded joint properties and metallurgy in terms of their application or possible application in mechanised, robotic or automated systems.

In a series of volumes they commence with such an outline and continue with texts which give more detail on the arc and laser fusion welding processes, discuss the characteristics of machines for welding, detail the operation and example the types of robots which can handle and perform fusion welding, and then consider the needs and methods by which the processes can be adaptively controlled. Finally, the way in which the various elements (robots, process, adaptivity, cell control and management, etc.) are linked together is described. In editing the English translation of these volumes the original content and intent of the authors has, where possible, been retained. However, the differences in language, in national welding standards and terminology have in places resulted in a need to modify and add to the text.

This series, in leading gradually into the subject, will be of value to all those who need to consider the mechanisation of fusion welding processes, whether they be students, welding engineers, designers, managers, production engineers or robots application engineers. The early volumes should also appeal to those pursuing general studies of fusion welding.

This first text commences with chapters on the basics of fusion welding technologies and the definitions of processes and weld joints – foundations on which the series is built. In producing a welded joint, either by manual or machine methods, consideration must be given to the physical and mechanical properties of the joint and to the shape of the final structure. These topics and the pre- and post-weld treatments which can affect them are discussed in basic terms but always with the underlying consideration of the need to use them in an automatic situation.

John Weston, The Welding Institute
English Language Editor

CONTENTS

PREFACE

ALTHOUGH it is now self-evident to assert that without welding there would be no modern industry, it is still worth distinguishing two great families of processes in welding.

Archaic processes: With heat sources of modest and diffuse temperature, even at the most favourable temperatures the welding occurred over long periods and large areas were heated. Craftsmen in non-industrialised countries still work to this traditional formula. However, certain large and highly specialised industries exploit modernised forms of forge welding. For example, Fretz Moon pipe welding and manufacturing steel plates by rolling or explosive bonding.

Modern rapid weld processes: Using diverse concentrated heat sources resulting from relatively recent scientific discoveries at the beginning of this century. However, it has only been in the last 50 years that they have been subject to effective industrial development.

During this brief period of time and by a multiplicity of processes (the range in modern industry includes more than 100), welding has been introduced throughout manufacturing industry where metal is involved and has been one of the enabling technologies, be it the finest of electronic components or the heaviest and most powerful constructions. Welding has aided the development of products ranging from household equipment through the building industry to the construction of bridges, methods of transport (cycle, car, railway, marine engine, aeronautics), hydraulics, electrical and electro-mechanical constructions, heavy engines, presses, machines of all types (mining, handling and public utilities equipment, heavy machine tools, etc.), metallurgical, petrol, chemical and gas equipment, materials for transport and distribution by piping, heating industries, and finally to the most advanced nuclear and space technologies.

Without welding, almost all modern technical achievements would not exist. Welding simplifies assemblies and reduces their bulk, and which, when correctly carried out, guarantees the best strength for the minimum mass and volume, the perfect seal against air and water, and the best thermal, electric and magnetic continuity. As an example: an oil tanker weighing 270,000 tonnes, whose gross hull weight was 35,000 tonnes, requires 700 tonnes of welds, representing a weld bead length of the order of 650km.

With the exclusion of manual metal arc electrode welding, it is continuous electric arc welding, particularly with gas shielding, which has seen the greatest development in recent years. This is to a great extent thanks to the progress of automatic machines, to the appearance of digital control, and programmable automatic machines, etc. However, we can count, in the near future, on a new evolution in welding, in particular for large scale installations and through the spectacular use of lasers, particularly their strong concentration of energy.

Another factor determining the greater and more extensive use of welding techniques is linked to the vital need of industry to embrace automation. The appearance of robots and the new opportunities for automation which they bring, reinforce this acceleration effect: the industrial development of devices which can recognise shapes and a series of joints; the possibility of generating commands allowing modification of joint configuration in a single run, in multi-run welding or from one component to another; the use of computers to program robots; the use of increasingly powerful and rapid computers for controlling the different variables of the processes. All assist the genuine introduction of automatic welding in the widely different industries, increasing both the quality of the finished product as well as the rate of deposition. Further increases in deposition rate and in the areas involving welding can be expected.

One very significant fact is the importance of the welding research undertaken in all industrialised countries, especially as concerns the physics of the processes which is essential for the development of effective 'intelligent' robots with sensory capability. In fact, with manual systems, the operator can vary the control parameters according to what he sees and hears, and it is only necessary to know the direction of response of the different parameters and their basic controls. Welding automation, however, which assumes that the machine is capable of adapting itself to the uncertain fluctuations of the joint being welded, necessitates knowledge of the laws governing the action of the control parameters as well as a determination of the limits within which a joint can be made of sufficient quality.

The further evolution of welding assembly technology will profit from the numerous advances in fields as diverse as electronics, automation and metallurgy. Welding in the years to come will become more and more important and will be quite different from that which we know today. This will not be a brutal change but a continued evolution which becomes more and more rapid. If one wishes to predict, it could be said that manual welding with coated electrodes will be progressively replaced with processes using gas-shielded arc welding as well as with new processes such as the laser.

The future of the different processes will be greater the greater their potential for automation (and in particular for being performed by robot), as this will allow improvement in the quality of joints welded, and increase in productivity (e.g. increase in speed of welding, reduction or elimination of the need for added filler materials, etc.).

Technology and processes of welding. Robotisation of welding processes does not just involve the robot. Even when robots are well adapted for welding and are suitably set up, the users are still confronted by problems posed by the welding equipment, the preparation of assemblies and the process itself.

Today's robots, even if highly flexible (for machines), are not as universal as one would want. For example, a laser welding robot, whose beam is generated without the need to contact the metal being welded, will be very different from arc welding robots used today – notably in the speed of forming weld beads and the bulkiness of the pieces being assembled. In fact, history shows that the evolution of a technique is brought about by the continual interaction between the usable processes and the means of applying them.

On the other hand, the metal for an application (its nature, shape and preparation, etc.) affects, given the quality of the assembly to be obtained, all aspects of the welding operation, in particular those of processes (even with perfect mastery of parameters) and, therefore, in turn influences the choice of machine for the application.

It is, therefore, impossible to talk about automation or robotisation without a good knowledge of the welding and metallurgical rules to be followed and their interactions.

This first volume sets out to give a sufficient knowledge of the welding technologies and the various process developments to enable the reader to embark on an analysis of the condition or situations relating to their use.

Jean Cornu

Chapter One

HISTORICAL DEVELOPMENT OF WELDING

TO UNDERSTAND the evolution and development of welding, it is necessary to make a quick review of its history.

The term 'weld' (the joining of materials by welding) appeared for the first time in the Old Testament several thousand years BC. Indeed, 4,000 years ago the Egyptians had already developed the art of welding (the uniting of two or more parts by heat or pressure, or both). The head of the sarcophagus of Tutankhamen (1361–1352 BC), whose tomb was discovered in 1922 in the Valley of the Kings, is a good example. Another notable example from history is the famous Colossus of Rhodes (built c. 292–280 BC), in its time one of the Seven Wonders of the World, which owed its height of 35 metres to a skeleton of welded iron. Also, in the Ukraine about 1,200 years ago, craftsmen welded a magnificent steel sword which has survived to the present day.

The first societies who knew how to work metal only had available, for assembly, the rudimentary methods of flow and forge welding. They also practised brazing. These processes remained almost unchanged up to the middle of the nineteenth century. At this time, methods of shaping were progressively perfected, mechanical methods of assembly were developed and gradually replaced the primitive and often mediocre methods of welding.

Then, in the second half of the nineteenth century, the forge welding process was perfected and for specialised applications proved highly productive. Electric arc and resistance welding processes were invented in the same period. These constituted the first step towards the modern processes involving high energy densities, but they were not used to any notable degree due to the slow development of electricity and the rudimentary development of the actual process.

It was chemical processes which were the first to rejuvenate the art of welding: first thermit welding, a development of flow welding, whereby molten metal is poured (cast) into the joint area to melt the joint faces and to provide the filler material*; then came oxyacetylene welding, which, between 1905 and 1930, became the universal method of welding.

From 1925 onwards, the electric arc and resistance welding processes became established, gradually replacing oxyacetylene welding and allowing mass production and the production of large items.

Arc welding could not develop until there was some method of protecting the arc and molten weld pool from the atmosphere. The first solution, employed at the beginning of the twentieth century, was the covered electrode, where a metal electrode is covered with a coating whose essential role is to produce, through the heat of the arc, gases which would protect the molten weld metal from the atmosphere. Adding other substances to the coating influences metal transfer in the arc, the chemical composition of the deposited metal, its metallurgical structure, and hence its mechanical properties.

Furthermore, some of the constituents of the coating fuse to form a slag which on cooling cover the deposited metal, thus improving its appearance and protecting it from cooling too rapidly – rapid cooling carries a risk of weld brittleness (see Chapter Six).

In the manual metal arc process the welding current flows along the whole length of the electrode which limits its length (to a few hundred millimetres) because of resistance heating. This heat build-up can lead to deterioration of the coating. Resistance heating also restricts, for a given electrode diameter and length, the welding current that can be used.

This finite length of electrode leads or relates to the principle disadvantage of the process: its low productivity which results in part from the need to regularly change electrodes. To overcome this limitation, the concept of a continuously fed consumable electrode (e.g. a wire) was developed. Furthermore, when the welding current is fed to the consumable only a few millimetres above the arc, the resistance heating effect is minimised and higher currents and current densities can be used.

These differences are compared in Fig. 1.1 and below:

● Process (a):
  – manual metal arc electrode, 3.2mm diameter
  – current density, 12A/mm^2
  – utilisation factor 5–20%

---

* Thermit(e) is a mixture of finely powdered aluminium and iron oxide that produces a high temperature on combustion and the reaction products: steel and aluminium oxide. The steel can then be cast into the joint as described. Other metal/oxide combinations can be used for joining metals other than steel.

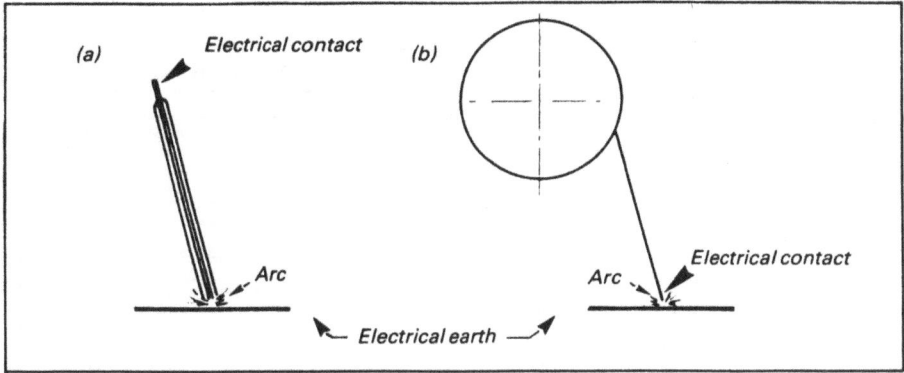

Fig. 1.1 Comparison of arc welding with: (a) basic covered and (b) continuous covered wire electrode

- Process (b):
    - continuous electrode wire, 1.2mm diameter
    - current density, $250A/mm^2$
    - utilisation factor, 50–90%

The developments and improvements which have been made to the covered electrode have led to the wide use of this arc welding process – commonly referred to as manual metal arc welding. It allows high quality welds to be made in most steels and remains the most commonly used arc welding process.

In other forms of electric arc welding, protection from the atmosphere is achieved in other ways:

- By a powder (flux) which completely covers the arc and the pool, e.g. submerged arc welding which was first used in the 1930s.
- By a gas which controls the atmosphere in the region of the arc and the weld pool, e.g. gas shielded metal arc welding (GSMAW) which found increasing use after the Second World War. When the shielding gas is inert (e.g. argon), the process is often known as metal inert gas (MIG) welding, and when an oxidising gas (e.g. carbon dioxide) is added it is known as metal active gas (MAG) welding. Unfortunately these terms (and carbon dioxide welding) are often used indiscriminately.
- By a flux (often of similar composition to that used to coat manual metal arc electrodes) contained within a continuous wire tube. These 'flux-cored' wires may be 'self-shielded' (the core of flux generating the protective gases in the heat of the arc) or may have auxiliary gas shielding added as for GSMAW. These consumables were first used in industry around 1958.

The principle stages in the development and introduction of welding processes are outlined in Fig. 1.2. As can be seen, the period since the mid-1930s has been particularly prolific. Since the invention of a basic flux-covered electrode in 1937, there has

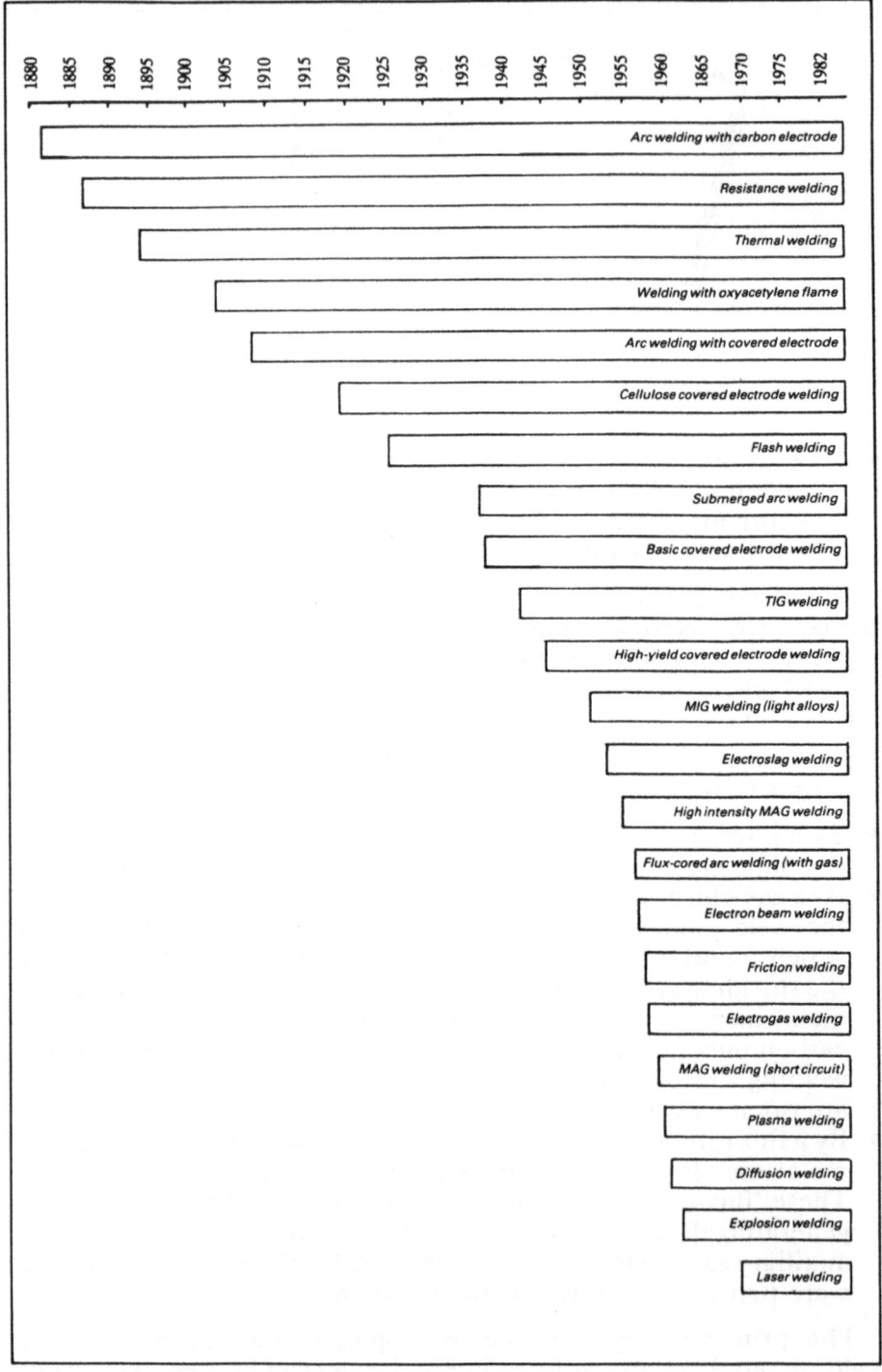

Fig. 1.2   Historical development of welding

followed, for example, MIG welding of light alloys (1949), MAG welding of high strength steels (1953), flux-cored arc welding with and without gas protection (1954), MAG welding with short-circuit transfer (1957), plasma welding (1958) and, most recently, laser welding (1970).

Submerged arc welding, which uses a continuously fed (wire) electrode and a layer of protective flux, was first used in industry in the USA in 1936, and it signalled the introduction of mechanised arc welding. Inert gases were used for the first time in 1941 in the USA as a protective atmosphere for tungsten inert gas (TIG) welding (also referred to as gas tungsten arc welding, GTAW) of light alloys in the aeronautic industry. MIG welding appeared on the market in 1948 in the USA (1952 in Europe) for semi-automatic (the electrode wire fed 'automatically' and the torch movement manual) welding of aluminium alloys.

It was demonstrated in 1953 that carbon dioxide could be used as a protective atmosphere for the welding of steel, given that a wire containing sufficient deoxidising elements was used. In 1957, MIG welding using a short-circuiting transfer technique was developed independently in the USSR, the USA and the UK. This variant which operates with a lower current density and a lower voltage, revolutionised the welding of sheet metal in mass production and automotive industries. The small, rapidly solidifying weld pool produced enabled the welder to make continuous precision welds in all positions. These processes, because they can operate without interruption, can and did become part of mechanised equipment. A MIG welding torch was, in 1962, the 'accessory' added to a machine tool for the machine welding of castings.

Even though the basic concepts of all those welding processes in current use had been outlined before 1960, they have, since that time, experienced considerable development. They have become easier to use, have greater reliability and are capable of producing higher quality joints. New techniques and process variants have continued to evolve and at present the number probably exceeds 100.

The welding processes are practically always competitive with each other and it is very difficult to gain an exact idea of their distribution, especially as this differs from one industry to another. Nevertheless, Fig. 1.3, by showing the relative consumption of welding consumables, indicates the main trends in France between 1970 and 1982. Similar data from the USA is presented in Fig. 1.4.

Whereas the proportion of submerged arc welding has remained almost constant, there has been an increase in the use of MIG/MAG processes at the expense of MMA electrodes which has continued during subsequent years (i.e. the proportion of MIG/MAG consumables used per year is greater than the MMA electrode usage).

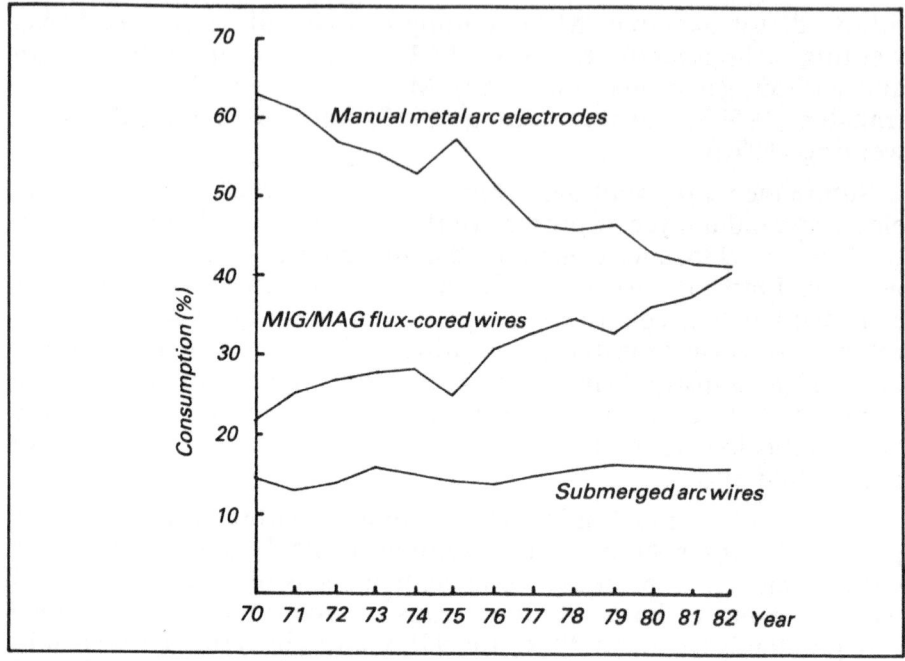

Fig. 1.3   Electrode consumption in France, 1970–82

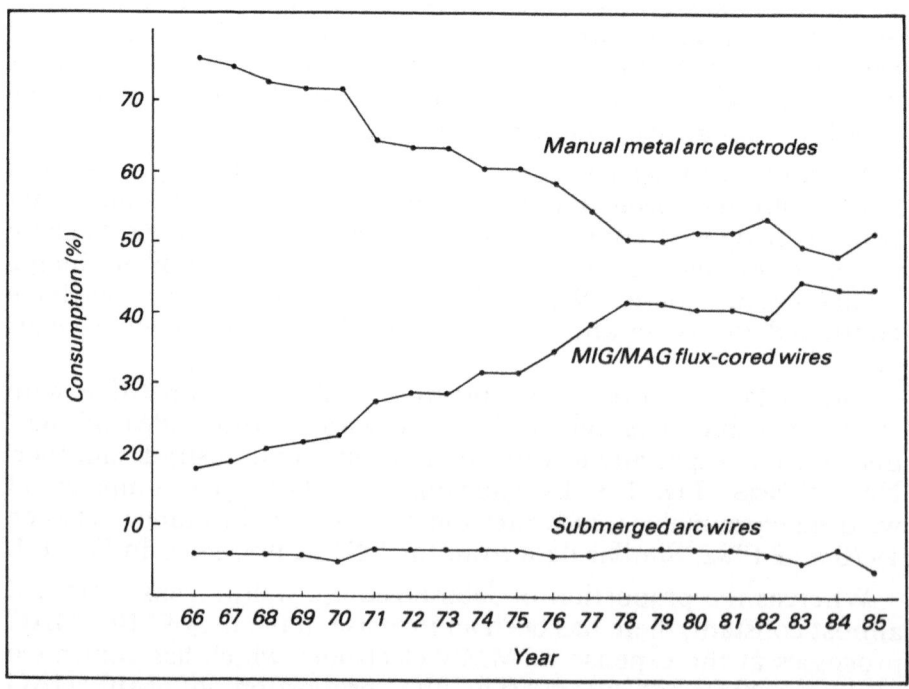

Fig. 1.4   Electrode consumption in the USA, 1966–85

Practically this same situation also exists in the USA and West Germany. In Japan, MIG/MAG welding, which was at first only slowly exploited, is now undergoing extremely rapid expansion and its use should soon overtake MMA electrode use.

The increasing importance of the TIG and plasma processes is in part reflected by the increase in the use of argon for welding (Fig. 1.5). However, the MIG/MAG processes also use argon, or argon-based gas mixtures, and while their rate of use for weld pool shielding is high, about twice that of the TIG process, this is somewhat balanced by the lower productivity of the latter (between 50 and 60% of MIG/MAG).

The first demonstration of flux-cored arc welding took place at the American Welding Society (AWS) exhibition in Buffalo, New York, in 1954. This process, as mentioned above, began to be used industrially in 1958–59. Its development and application in the USA was rapid and by 1981 represented more than 17% of metal deposited, and a percentage of 20% was expected by 1985. This process is also being increasingly used in the rest of the world.

An important stage in the development of the arc process for positional welding was reached with the development of the power source able to pulse the arc at mains-related frequencies (e.g. 50, 100 or 150Hz). It is certain that higher frequency pulsing, now possible, which among other things can be adjusted to give one metal drop transfer per pulse (see Chapter Two), will bring further benefits.

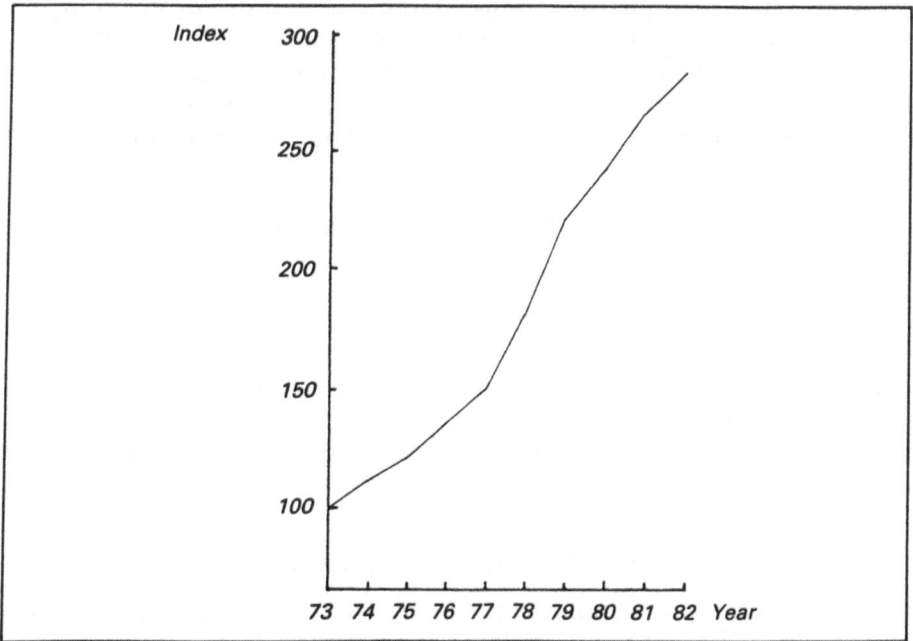

Fig. 1.5   Use of argon in welding in France, 1973–82 (base index: 100 in 1973)

The use of such pulsing which can be used to control the frequency of transfer and the shape of the weld bead, is extremely attractive in terms of minimising spatter and of modifying penetration while not altering the mean current. This technique can be applied equally well to both solid and flux-cored wire welding. With both techniques it is necessary to use a shielding gas which allows an axial mode of metal drop transfer (e.g. argon, argon-oxygen mixtures or argon-carbon dioxide mixtures containing less than 20% carbon dioxide).

Sinusoidal pulsing, because of the relatively slow rates of change of current, is not as good for welding as rectangular waveform pulsing. Modern electronics have made it possible to produce high-frequency rectangular waveform pulses at reasonable cost. The most recent welding power sources use invertor technology which has given increased control flexibility and has enabled a significant reduction in size to be achieved.

This control has enabled the development of what has become known as 'synergic welding', where the rate of feeding the consumable wire is linked automatically to the pulsing parameters, e.g. the frequency and the arc voltage. In effect this gives 'one-knob' control of the MIG/MAG processes.

The invertor power sources, which are easy to control, reduce spatter and can increase welding speeds, are likely to have a bright future in automated welding despite their cost – considerably higher than more traditional generators (approximately five-fold).

Consideration should also be given to the most recently developed welding process – laser welding – which because of the narrow fusion zone minimises distortion and keeps overheating of the parts (and the environment) to a minimum. Where welding times are short and solidification rapid, the welded assembly remains practically at the ambient temperature.

The use of lasers in the metalworking industries can be summarised as follows:

- The laser beam can be easily focused. It can be transmitted for several metres through the air without its characteristics being altered, and then focused for the particular task.

- The laser beam can be time-shared between several workstations or can be split into several independent beams for carrying out different functions simultaneously.

- Welds in steels up to 20mm thick are possible.

- It is possible to cleanly cut thicknesses exceeding 30mm in normally difficult to cut materials such as the alloys of nickel and titanium.

- Lasers can be used to heat or surface treat, at a great speed, surfaces of complex shape and contour without distortion.

- Lasers can be used for surfacing. Usually an alloy powder on the component surface is fused-in by the laser beam – the alloy elements being chosen or adjusted to withstand the particular surface characteristics required (wear, corrosion, etc.).

The above list indicates the potential range of applications for lasers. Despite the relatively large capital investment which the purchase of laser equipment represents, it has been established that for certain applications this cost can be economically justified. Lasers have many advantages: their excellent performance and their particular suitability for automation make the prospects for lasers in the working of metals seem assured.

However carefully an examination is made of the evolution of welding processes over the past few years, it remains impossible to predict exactly in what way welding will evolve in the future. It remains necessary to consider the great technological changes presently occurring in the welding environment. The future of those welding processes which can be mechanised appears brightest in that they facilitate:

- An increase in productivity through increased welding speed, reduced downtimes, and elimination or reduction of preparation angles.

- Improvement in the quality of welded joints.

- Adaptation to automation, particularly their use by robots.

Chapter Two

BASIC DEFINITIONS IN WELDING

IN ANY subject it is important, for two-way understanding, to know and to use the appropriate vocabulary. Welding is no exception to this rule. This chapter therefore presents a few basic definitions.

A *weld* is a permanent and continuous bond between two or more elements of solid material, made by welding.

Welding is an operation in which two or more parts are united by means of heat or pressure or both in such a way that there is continuity in the nature of the material (metal) between these parts. A filler metal, whose melting temperature is of the same order as that of the parent material, may or may not be used. Although most solid materials can be welded, this discussion is limited to the welding of metals and alloys.

Fusion welding is a class of welding in which the weld is made between surfaces brought to the molten state, with or without the addition of filler metal, and without the application of pressure. For fusion welding it is necessary to have a source of heat which can be directed at the joint area and which is sufficiently intense to produce rapid fusion of the metal (and filler) to be joined (steel melts at about 1500°C). The energy (heat) sources most often used are: flame, electron bombardment, light (laser) and, above all, the electric arc. Processes falling within this class include: arc welding, gas welding, aluminothermic welding, electron beam welding, electroslag welding, and light radiation (e.g. laser) welding. Fig. 2.1 shows the elements of a typical V-prepared, fusion weld made between butting plates.

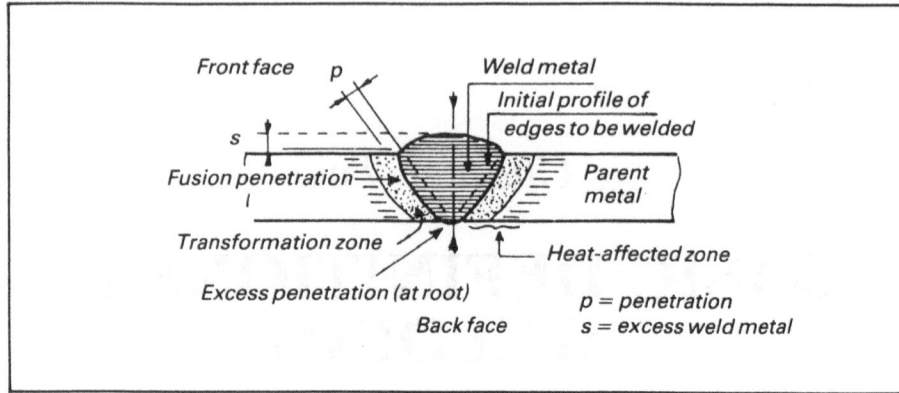

Fig. 2.1   Butt joint weld (V-shaped preparation)

Pressure welding (solid-phase welding) is a class of welding in which a weld is made by the application of sufficient pressure to cause plastic flow (or solid-state diffusion) of the surfaces, which may or may not be heated.   Processes falling within this class include: resistance (spot) welding, forge welding, pressure welding, ultrasonic welding, diffusion welding, and magnetically impelled arc butt welding (MIAB).   With most of these processes the equipment is sizeable and there is a need to grip or clamp the parts to enable the forging force to be exerted.   These two constraints, which limit the dimensions of parts which can be welded, seldom apply to the fusion welding processes.

Brazing is a joining class, different from welding.   The process is generally applied to metals in which, during or after heating, molten filler metal is drawn into or retained in the space between the closely adjacent surfaces of the parts to be joined by capillary attraction (Fig. 2.2).   In general, the melting point of the filler metal is above 450°C, but is always below the melting temperature of the parent metal.

Braze welding is a subset of the above class and is defined as the joining of metals using a technique similar to fusion welding and a filler metal with a lower melting point than the parent metal, but neither using capillary action (as in brazing) nor intentionally melting (as with fusion welding) the parent metal (Fig. 2.3).

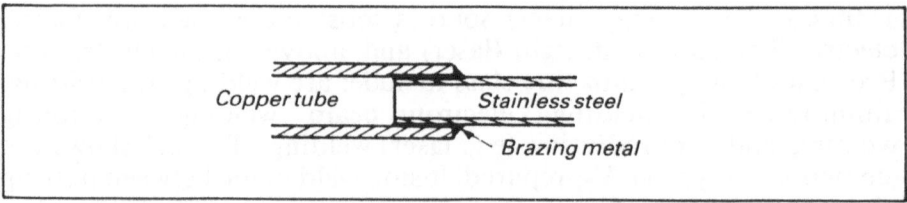

Fig. 2. 2 Brazing

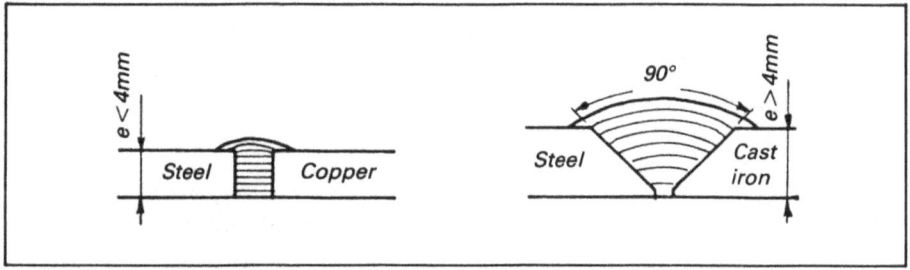

Fig. 2.3 Braze weld

Surfacing is the deposition of metal on a surface to provide a layer giving properties different from those of the parent metal.

Parent (base) metal is the metal to be joined or surfaced by a welding or brazing process. Parent materials may be dissimilar.

Filler metal is the metal added during welding, brazing or joining.

Deposited metal is the filler metal after it becomes part of a weld or joint.

Fusion zone is that part of the parent metal that has been melted into weld metal.

Fusion penetration is the depth to which the parent metal has been fused (often simply 'penetration'), (see Fig. 2.1).

Weld junction is the boundary between the fusion zone and the heat- affected zone.

Joint is a connection where the individual components, suitably prepared and assembled, are joined by welding or brazing, e.g. butt or fillet.

Molten pool (weld pool) is the pool of liquid metal formed during fusion welding (in electroslag welding the slag bath is included).

Weld metal is all the metal melted during the making of a weld and retained in the weld (see Fig. 2.1).

Heat-affected zone (HAZ) is the part of the parent metal that has been metallurgically affected by the heat of welding (or cutting) but not melted (see Fig. 2.1). For those metals which have an allotropic transformation (during cooling from near their melting point), there will be, within the HAZ, a zone known as the 'transformation zone' where this microstructural transformation has occurred. In the remaining part of the HAZ there is no allotropic transformation.

Run (pass) is the metal melted or deposited during one passage of the welding electrode or torch.

Bead is a single run of weld metal.

Root (of preparation) is the zone in the region of, and including, the gap where the first run is to be placed in a joint.

Root run (root pass) is the first run deposited in the root of a multi-run weld (Fig. 2.4).

Weaving is transverse (to the direction of welding) oscillation of a electrode or torch during the deposition of weld metal.

Stringer bead is a run of weld metal made with little or no weaving motion. Multi-run welds can be built up using either stringer runs and or weaving runs.

Sealing run (backing run) is the final run deposited on the root side of a fusion weld.

Backing is material placed at the root of a weld to contain or support the molten pool. Backings may be 'permanent' if they are penetrated by weld metal and become part of the joint (Fig. 2.5), or are 'temporary' if it is not intended that they become part of the weld.

Face (weld face) is the surface of a fusion weld, or the side of a fusion weld, from which the weld has been made.

Reverse (face) is the opposite side of the material being welded to the weld face.

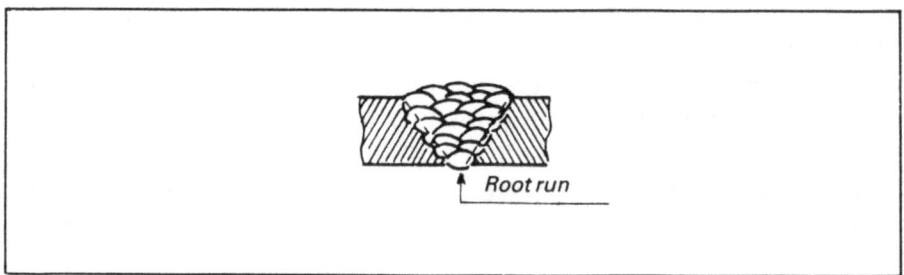

Fig. 2.4   Multi-pass weld

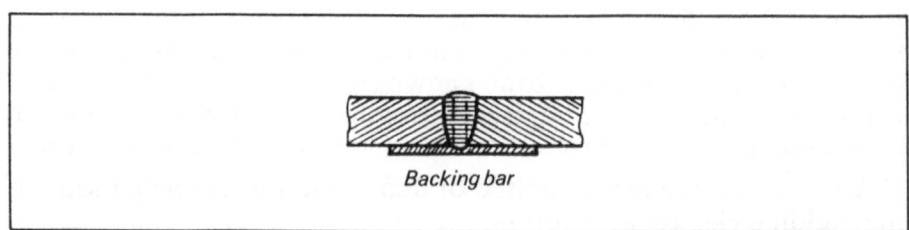

Fig. 2.5   Weld with a support on the reverse

Chapter Three

WELDING ASSEMBLY TECHNOLOGY

IN THE context of the use and working of metals, such terms as operating technique and welding procedure, assembly, joint, homogeneous and heterogeneous are often used. It is therefore desirable to clarify and define these terms.

The welding method or the *operating technique* refers usually to the method(s) or arrangement(s) adopted for welding an assembly (e.g. multi-pass welding, welding by balanced passes, groove welding, welding of successive blocks, welding with backing support, etc.).

The term *welding procedure* is more general. It includes the material condition and the preparatory operations (e.g. edge preparation and preheating), the welding method and the parameters (e.g. MAG welding with values for arc current arc current, voltage, travel speed, etc.), and the relevant environmental factors (e.g. temperature, air currents, etc.).

Normally, care is taken in the usage of the terms 'fabrication' (assembly) and 'joint'. A *joint* is limited to that proportion of material constituting the connection where the individual components, suitably prepared and fitted together, are to be joined by welding or brazing. The word *assembly* (or fabrication) defines the total collection of parts and includes the joint(s) which unite them.

It is useful to consider the usage of the adjectives 'homogeneous' and 'heterogeneous'. The expression *homogeneous fabrication* is a synonym for a fabrication of similar metal parts, and *heterogeneous fabrication* denotes a fabrication of dissimilar metal parts. On the other hand, a weld or a welded joint is homogeneous when the filler metal is identical or similar to the base metal. In the opposite case the weld is heterogeneous.

Some useful remarks are as follows:

- To eliminate all confusion, the expression 'weld with similar filler metal' and 'weld with dissimilar filler metal' are preferable to 'homogeneous weld' and 'heterogeneous weld'.
- When dissimilar metals are joined, the joint is heterogeneous.
- Brasures and braze welds are always heterogeneous joints, but assemblies may be homogeneous.

Welds, joints and fabrications

In producing a fabrication, it becomes necessary to weld together joints which are in various attitudes in space. These types of joints are defined in various standards (e.g. BS449) and are summarised below. First we should consider the form of material itself and in particular the grain flow, as this can influence selection of materials and joint types.

Grain flow effects

During rolling or forging processes the grains and inclusions within the metal are elongated and have, in metallurgical sections, the appearance of fibres or laminations. These orientation effects can result in the metal having different properties in different directions, especially in the through-thickness direction in rolled metals.

These property differences can have important consequences for the designer and welder, as the contraction stress, built up as a weld cools, can be sufficient to delaminate (i.e. crack) the parent plate material. Joints designed for joining end-grain to end-grain are less susceptible to this effect than are joints made end-grain to plate surface (e.g. T-joints).

Types of weld

There are four basic types of weld:

- *Butt welds* – in essence welds which have significant penetration through the material thickness (Fig. 3.1a).

- *Plug welds* – made by filling a hole in one component workpiece with filler metal, thus forming a joint with the underlying metal (Fig. 3.1b).

- *Edge welds* – deposited on the edge of metals which have their faces adjacent (Fig. 3.1c).

- *Fillet welds* – fusion welds, other than butt, plug or edge, which are approximately triangular in cross-section (Fig. 3.1d).

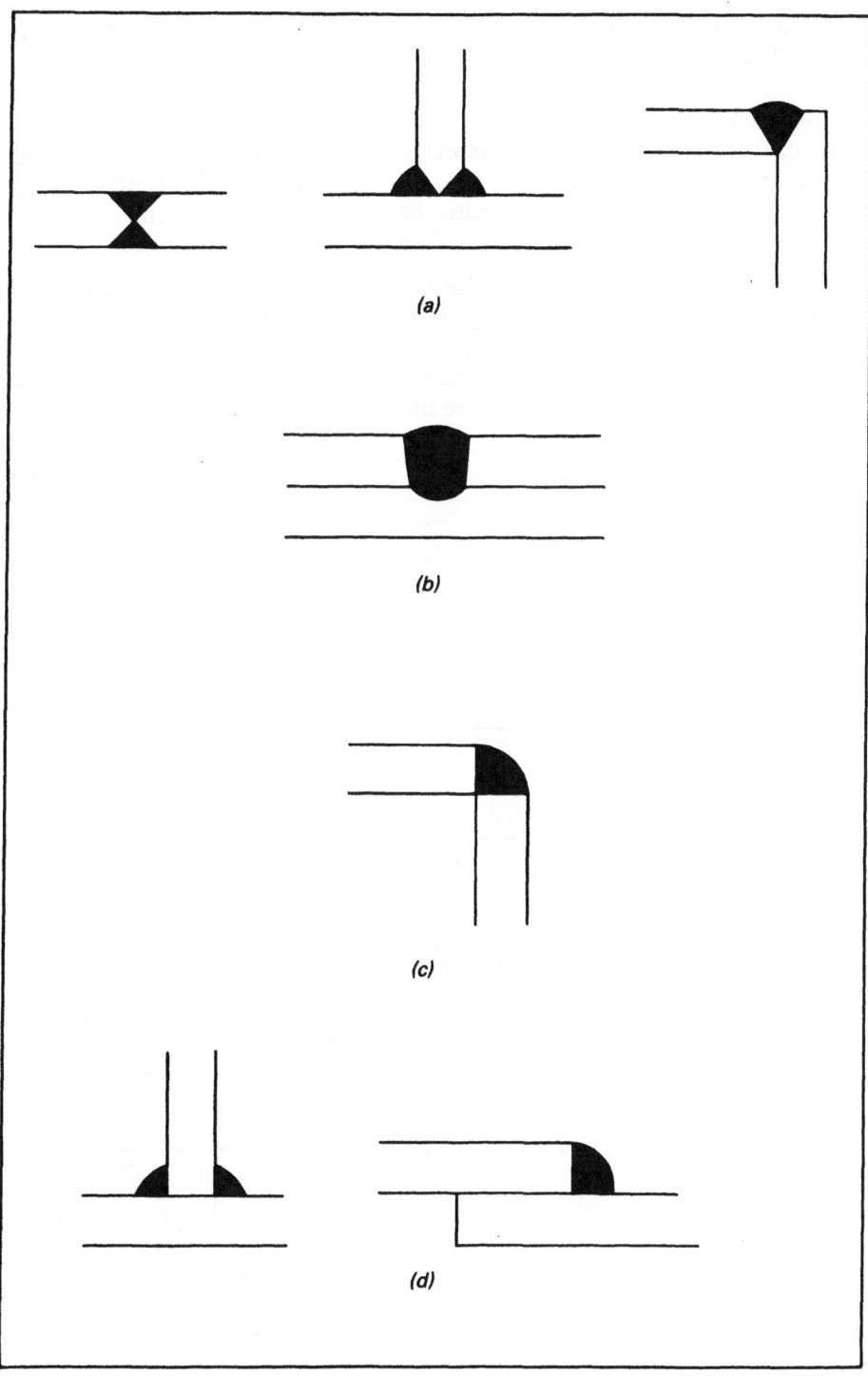

Fig. 3.1 Types of weld: (a) butt, (b) plug, (c) edge, and (d) fillet

These types of weld are used as appropriate in making the joints described below.

Types of joint
The categories of joint follow from the preparations, dimensions and relative orientations of the parts being joined. A consideration of a section through the joint, whether between square, circular, plate or tubular materials, enables the following types to be defined:

- *Butt joint* – a joint where the ends of the sheets or plates to be welded butt one another (Fig. 3.2a). Grain flow alignment is good. Butt joints are used whenever possible as they have the best properties in terms of resistance to tensile, compressive or fatigue loading. They can however be more difficult to produce than, for example, fillet welds.

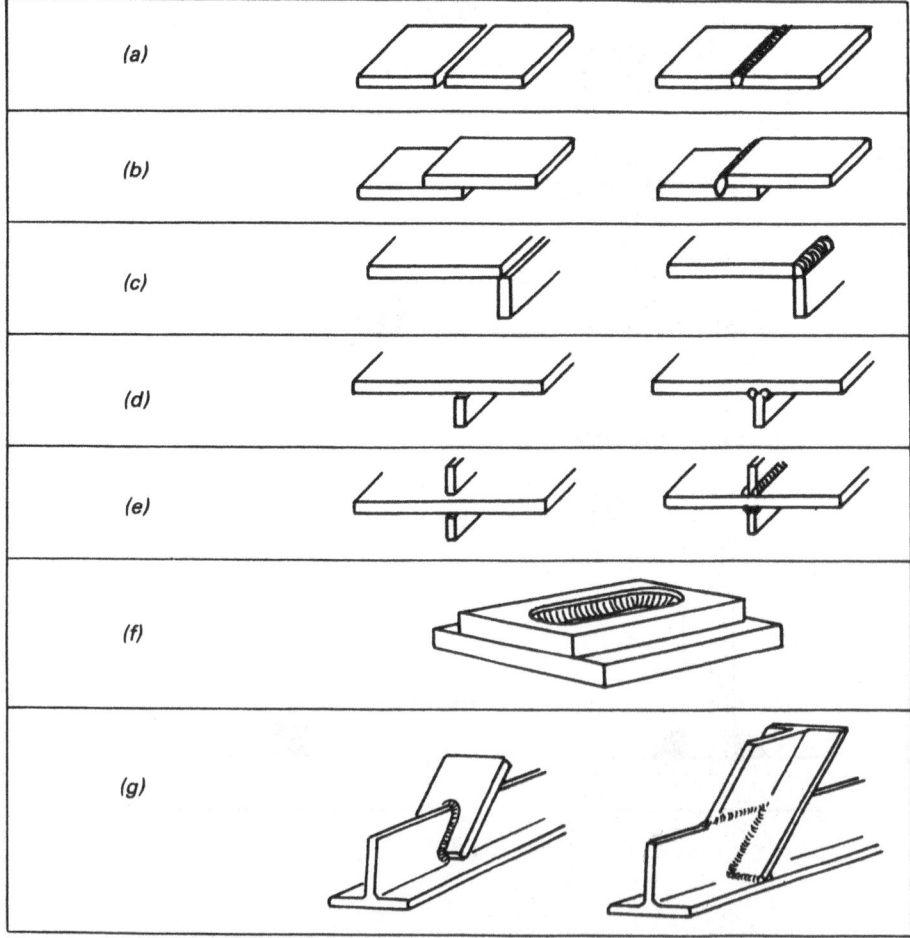

Fig. 3.2   Basic joint types: (a) butt, (b) lap, (c) edge, (d) 'T', (e) cruciform, (f) slot overlap, and (g) fork

- *Lap joint* – a connection between two overlapping parts (Fig. 3.2b). Lap joints are widely used to join sheet products. They can be easy to fit-up and weld but their strength is lower than the butt joint. Grain flow alignment is good on one plate but poor on the other.

- *Corner joint* – a connection between the edges of two parts making an angle to one another of between 30° and 135° in the region of the joint (Fig. 3.2c). Grain alignment varies from good to poor depending on whether the weld is of the butt or fillet type.

- *T-joint* – a connection between the edge of one part and the face of another which may be either butt or fillet welded (Fig. 3.2d). Grain flow in one plate is normal to that in the other – a situation that can lead to laminar defects when welding thick or sensitive materials.

- *Cruciform joint* – a connection in which two flat plates, or bars, are welded to another flat plate at right angles and on the same axis (Fig. 3.2e). Again the welds may be either fillet or butt. Grain flow alignment is again poor, and as residual stress levels can be high, joints of this type are those most susceptible to lamellar tearing. A design change, whereby a forged cruciform is butt welded to the surrounding assembly, might be used to avoid the problem.

- *Slot lap joint* – a form of lap joint where the joint is made by depositing a fillet weld around the periphery of a hole in one component so as to join it to the surface of the other component (Fig. 3.2f). The hole is not normally filled. Grain flow alignment is poor on one plate and this form of joint is susceptible to distortion. Standards often place limits on the size of slots. For example, the slot length must not exceed 10 times the plate thickness, the radius at the ends of the slot must not be less than plate thickness and the slot spacing must be more than four times the slot width.

- *Plug joint* – a form of slot lap joint in which the hole is circular and during welding is filled with weld metal. It is usually used between sheet materials of less than 10mm thickness. Standards again place restrictions on hole size, (e.g. diameter must not be less than sheet thickness plus 8mm nor greater than 2.25 times sheet thickness).

- *Fork joint* – with these joints (Fig. 3.2g) there are problems associated with the achievement of a satisfactory grain flow alignment, the production of the welds (or adjacent welds) because of restricted access, the avoidance of details which involve eccentric attachments, and the production (i.e. cutting) of the joint shape. Nevertheless, they can have advantages in aiding preweld assembly.

Welding attitudes

There are four fundamental welding attitudes or positions: flat, horizontal/vertical (H/V), vertical, and overhead, with the term 'inclined position' used with qualification to describe welding in other attitudes (Fig. 3.3).

These attitudes describe the way the bead lies in space. Whether butt or fillet welds are being discussed, the weld bead can be considered as a triangular prism of fused metal comprising a root, a cap or free face and two fusion faces (Fig. 3.4). The spatial attitude of this bead is then defined in terms of the angle of rotation α between the vertical plane and the plane bisecting the fusion faces (Fig. 3.5), and the inclination angle β between the line of the root and the horizontal (Fig. 3.6). The limits of these angles for each of the welding attitudes are shown in Fig. 3.7.

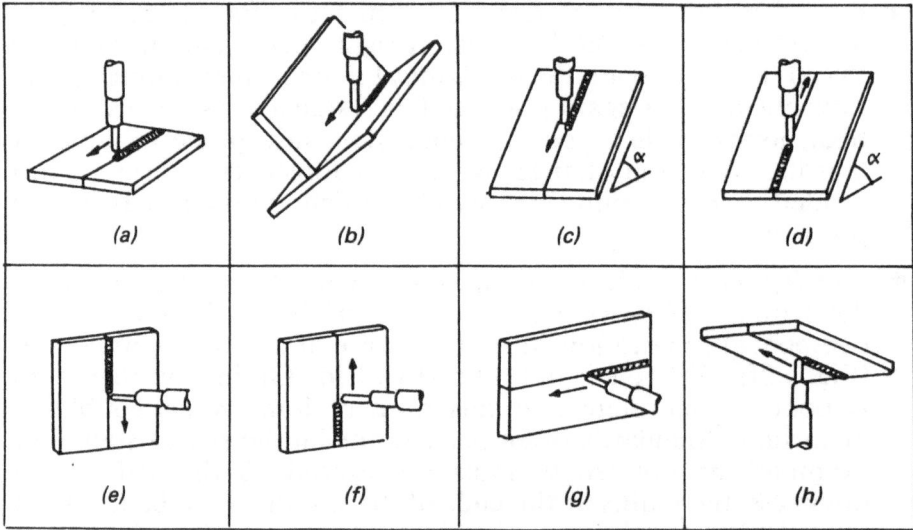

Fig. 3.3 *Welding positions: (a) flat, (b) flat fillet, (c) inclined (down), (d) inclined (up), (e) vertical (down), (f) vertical (up), (g) horizontal/vertical, and (h) overhead*

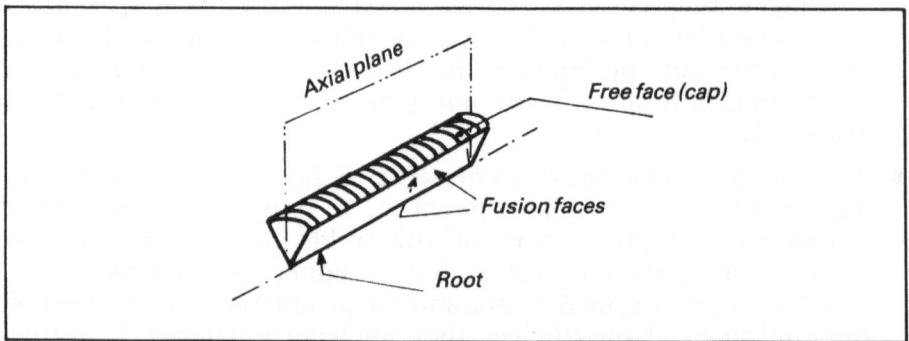

Fig. 3.4 *Weld bead formed by a prism of fused metal*

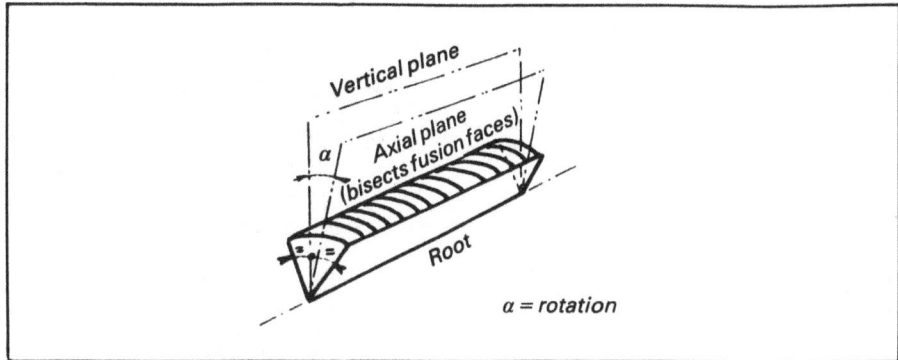

Fig. 3.5   Angle of rotation α

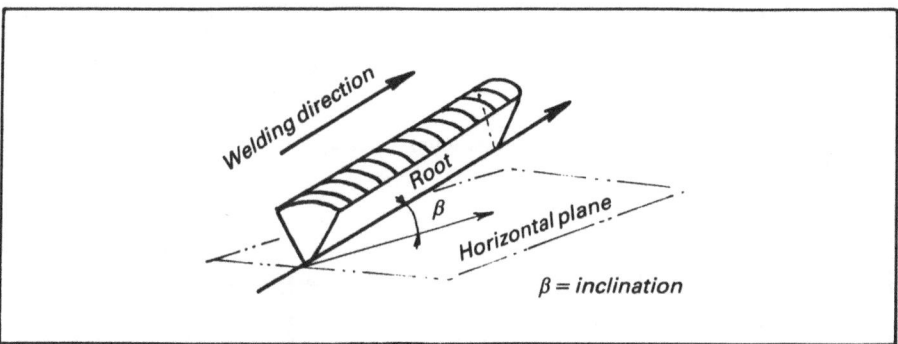

Fig. 3.6   Angle of inclination β

Joint preparations

With reference to Fig. 3.8, the joints, their preparations and the welding positions may be described as follows:

1. Prepared butt joint butt welded overhead.
2. Prepared butt joint butt welded flat.
3. Prepared butt joint butt welded H/V.
4. Prepared butt joint butt welded inclined H/V.
5. Prepared butt joint butt welded vertically up.
6. Prepared butt joint butt welded vertically down.
7 and 8. Double-side close (unprepared) butt joints welded both flat and overhead.
9. Lap joint fillet welded overhead.
10. Lap joint fillet welded H/V.
11 and 13. Fillet joints fillet welded H/V.
12. Fillet joint fillet welded overhead.
14. Edge joint edge welded H/V.
15. Slot lap joint fillet welded H/V.
16. Fillet joint fillet welded flat.
17. Edge joint edge welded flat.

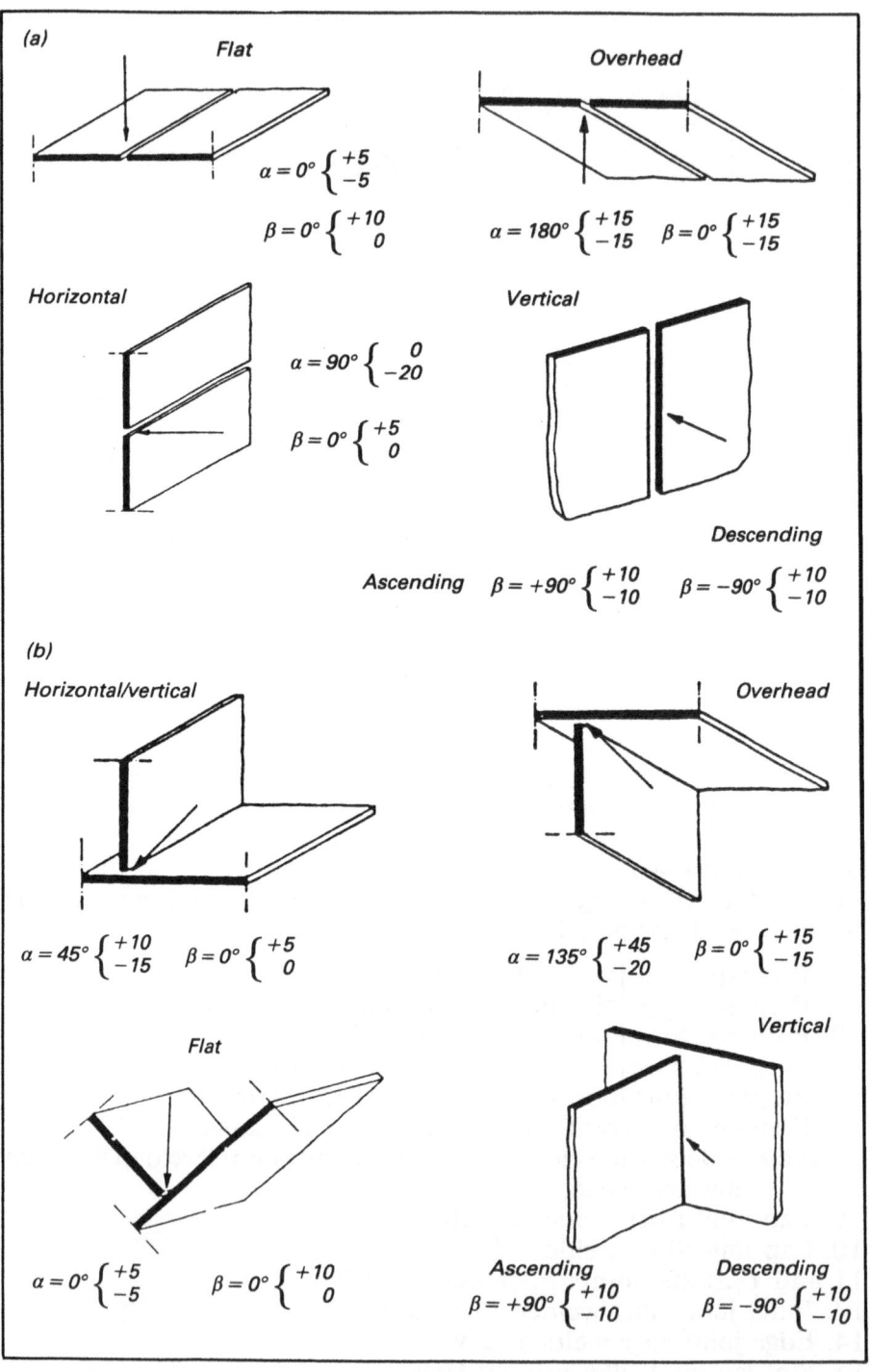

Fig. 3.7   The four fundamental welding positions for (a) butt and (b) fillet joints (after BS449)

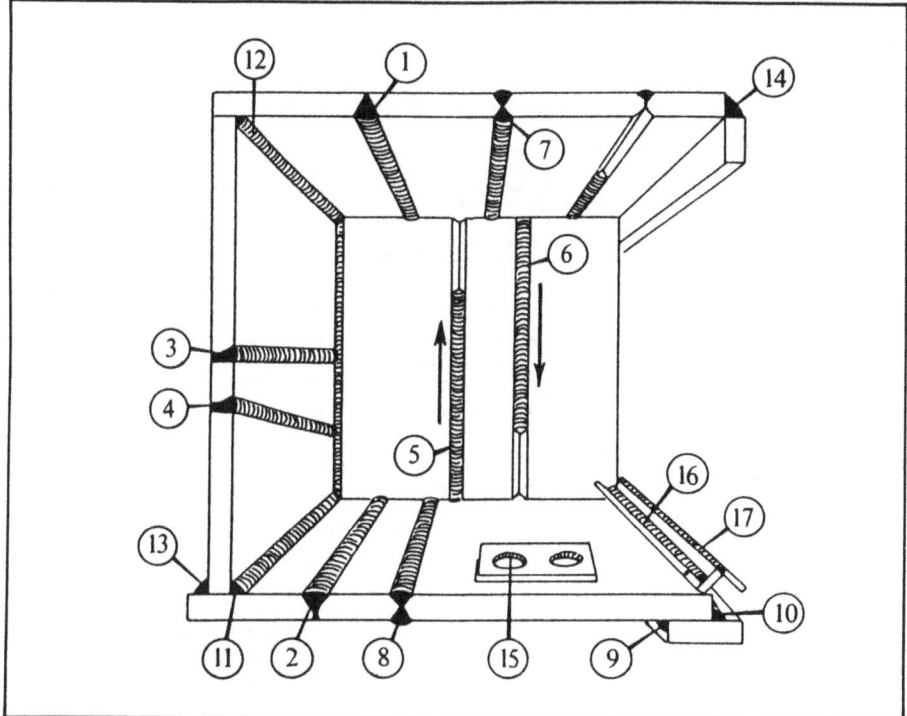

Fig. 3.8   Joint designation

Examples of the joint types and preparations given in Fig. 3.9 cover: thin sheet-to-sheet butt and lap joints, butt joints in plate, dissimilar thickness joints, T-joints, tube joints, and tube-to-flange joints.

(a)

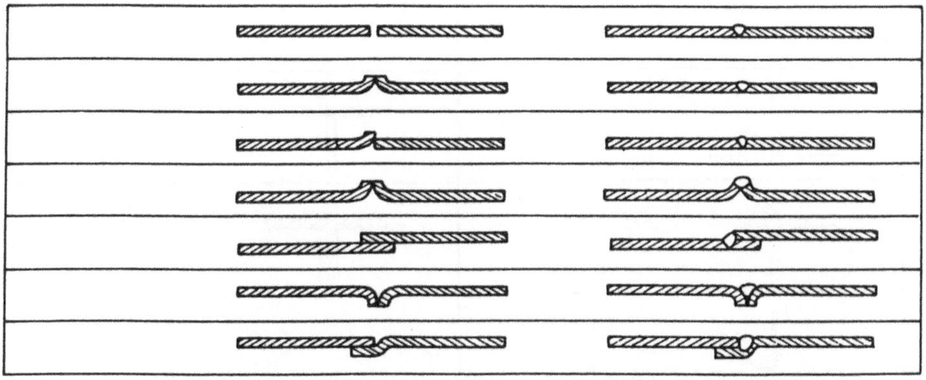

Fig. 3.9   Examples of joint preparations: (a) thin sheet butt and lap joints, (b) butt joints in plates, (c) dissimilar thicknesses, (d) T-joints, (e) tube joints, and (f) tube-to-flange joints (continued overleaf).

(b)

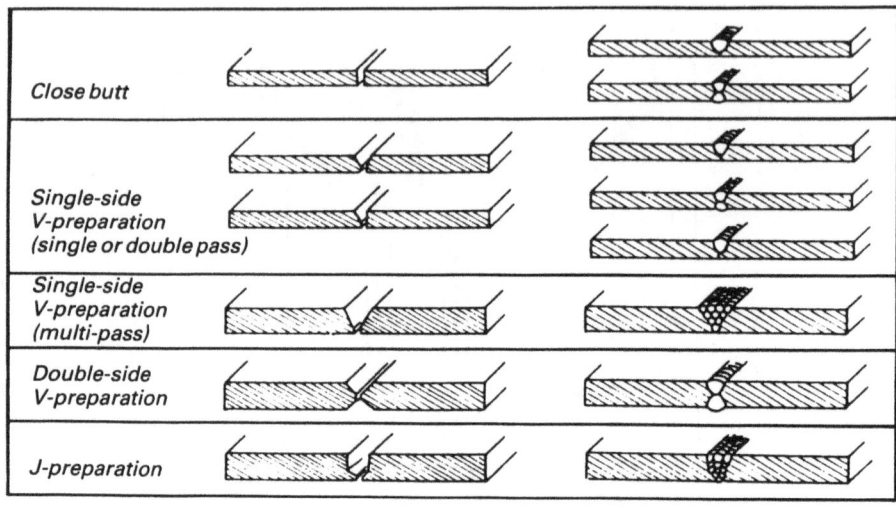

Close butt	
Single-side V-preparation (single or double pass)	
Single-side V-preparation (multi-pass)	
Double-side V-preparation	
J-preparation	

(c)

(d)

Fig. 3.9 continued.

(e)

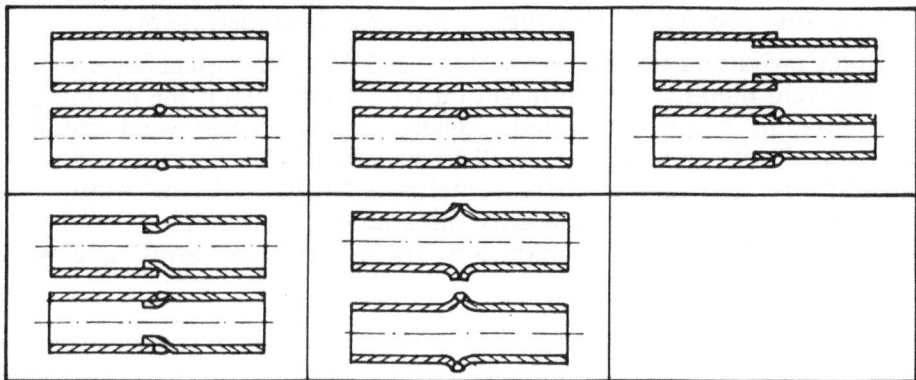

(f)

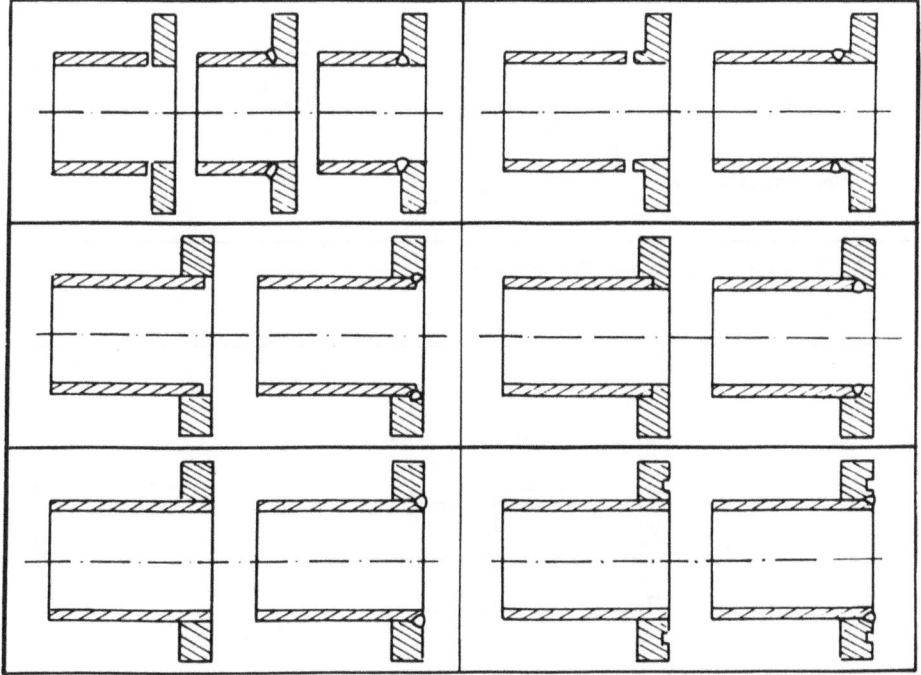

Fig. 3.9 continued.

Weld geometry

As described above, welds are largely classified within the two basic types of butt and fillet (Fig. 3.10). The significant and characteristic dimensions and features of these welds are described below.

In the cross-section of a butt weld (Fig. 3.11), the typical feature of excess weld metal s_1, which might only be acceptable if the height is restricted to b, can be noted. There is also excess weld metal associated with the root s_1. If all this surface metal is removed by grinding or machining, there remains a 'flush' weld of nominal thickness c, the 'design (throat) thickness'. The angles α and β represent the original preparation 'included' and 'bevel' angles, respectively.

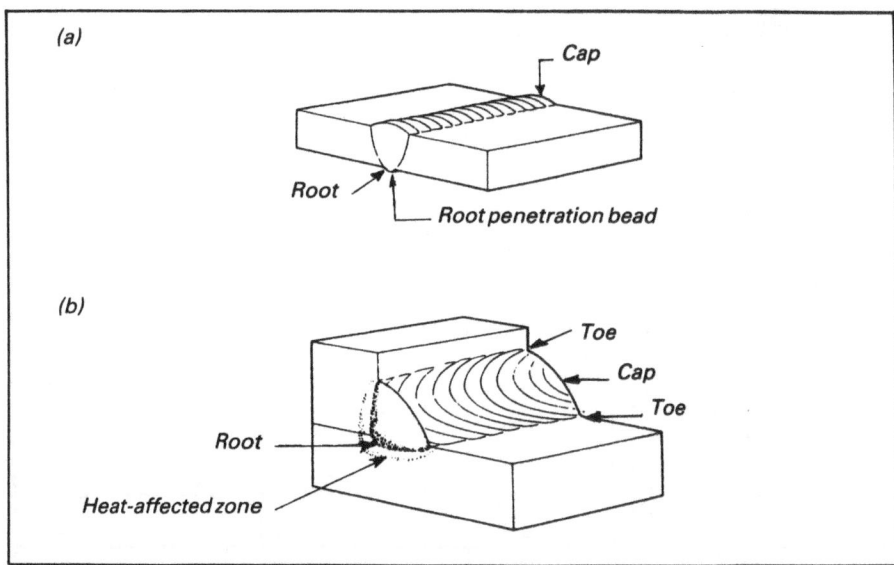

Fig. 3.10   (a) Butt weld and (b) fillet weld

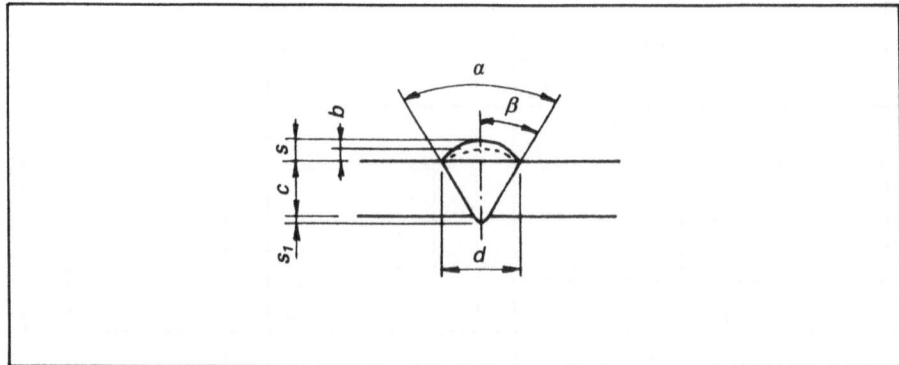

Fig. 3.11   Butt weld cross-section

A fillet weld may exhibit three forms of surface profile: flat, convex or concave (Fig. 3.12). Again there is excess weld metal s, 'design throat thickness' a and root penetration b.

The design throat thickness is, in the case of the flat and convex-shaped fillets, approximately related to the fillet leg length h by a factor of $1\sqrt{2}$. Thus with welds of this shape, the leg length can be measured to give an indication of throat thickness. With concave fillets the throat thickness itself can be measured. Such measurements do not take account of strength benefits arising from penetration.

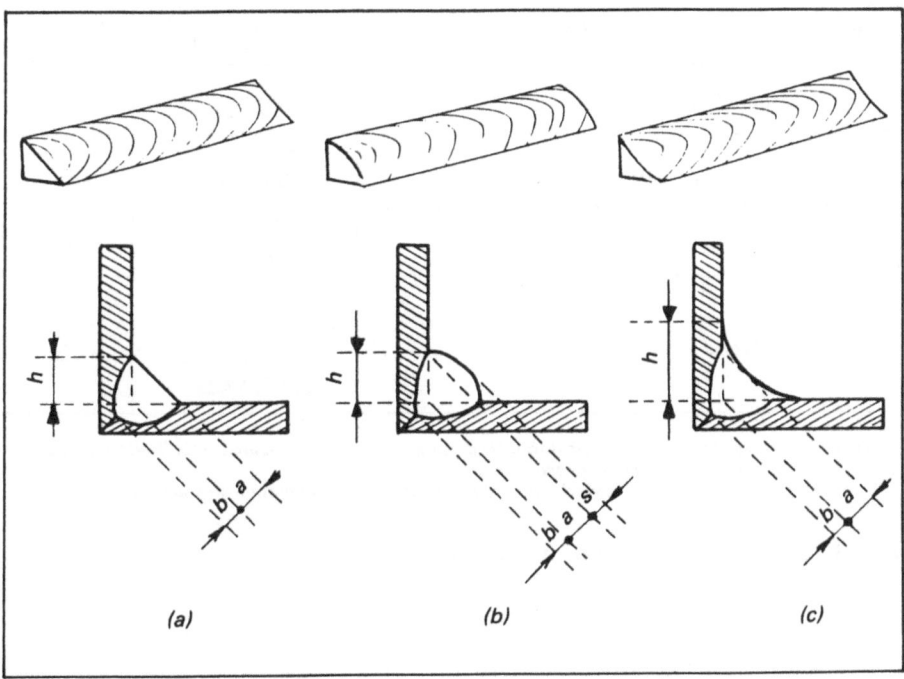

Fig. 3.12   Fillet weld surface profiles: (a) flat, (b) convex, and (c) concave

Welding symbols

Examples of welding terms and symbols are given in Fig. 3.13.

No.	Designation	Illustration	Symbol
1	But weld between flanged plates* (the flanges being melted down completely)		⊐⊏
2	Square butt weld†		II
3	Single-V butt weld		V
4	Single-bevel butt weld		V
5	Single-V butt weld with broad root face		Y
6	Single-bevel butt weld with broad root face		Y
7	Single-U butt weld		Y
8	Single-J butt weld		Y
9	Backing or sealing run		⌒
10	Fillet weld	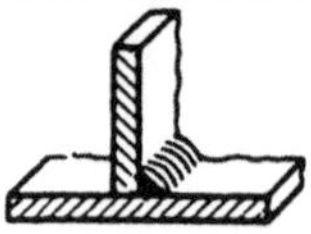	△

* Butt welds between flanged plates (symbol 1) not completely penetrated are symbolised as square butt welds (symbol 2) with the weld thicknesses shown.
† This symbol is used to indicate a stud weld when there is no end preparation and no fillet weld.

Fig. 3.13   Standardisation of welding symbols

Chapter Four

METALLURGICAL CONCEPTS

THE ADVANCED welding techniques are indebted as much to metallurgy as to other sciences. Throughout the ages man has conceived machines but has only been able to produce them much later when adequate materials for their construction had been developed. In many fields, notably aerospace, the progress made during the past 50 years could not have occurred without the development of new alloys. Today metallurgists and other researchers are striving to resolve current problems and to forge ahead of current needs through the development of new materials.

However, without an appropriate joining technique, the highly developed or prepared material is scarcely more useful than joinable material which has inadequate properties. Thus the welding metallurgist is faced with two problems: the improvement of the quality and reliability of welds made using traditional metals and alloys, and the development and/or adaptation of processes and techniques for the welding of new materials, the number of which is increasing all the time.

Surveys show that of the 800 groups of alloys known, a large number cannot yet be welded in a satisfactory manner. Furthermore, in the future there is likely to be available steels with tensile strengths higher than those currently available (and there will be a similar trend with the light alloys), which will pose considerable welding problems.

Mechanical properties of metals

Metals possess chemical properties inherent in their composition (e.g. resistance to corrosion), physical properties (e.g. density and thermal and electrical conductivity), and mechanical properties (e.g. tensile strength). The different properties are, however, usually linked. Thus, modifying the composition of an alloy to influence its resistance to corrosion can also alter its thermal conductivity and tensile strength. To gain an understanding of the mechanical properties and their relationships, some understanding of the atomic and metallurgical structure of metals is needed.

The metallic elements are made up of vast numbers of atoms which are arranged in orderly geometric patterns within the solid state (i.e. they are said to be crystalline). In each metal the atoms arrange themselves according to a specific geometric pattern called a lattice. There are many types of lattice, but most metals and their alloys crystallise in one of the following systems:

- *Body-centred cubic lattice* – the atoms, nine in total, are positioned at the corners and in the centre of a cube (Fig. 4.1a). Alpha iron (see Chapter Six), chromium and molybdenum crystallise according to this system.
- *Face-centred cubic lattice* – the atoms, 14 in total, are positioned at the corners of a cube and in the centre of each face (Fig. 4.1b). Gamma iron (see Chapter Six), copper, aluminium, nickel and manganese crystallise according to this system.
- *Hexagonal close-packed lattice* – the atoms, 17 in total, arrange themselves in a hexagonal prism with one atom at the centre of each end-face and with three internal atoms (Fig. 4.1c). This is the crystallisation system for magnesium and zinc.

The atoms are held in these lattice patterns by forces acting between the protons of the core and the surrounding electrons. It is these interatomic forces which give metals their properties, strength and rigidity.

As metals solidify from the molten state, a small group of atoms first cluster to form a crystal lattice nucleus. From this nucleus the crystal structure develops in different directions and the crystals grow until they meet each other. These irregularly shaped crystals,

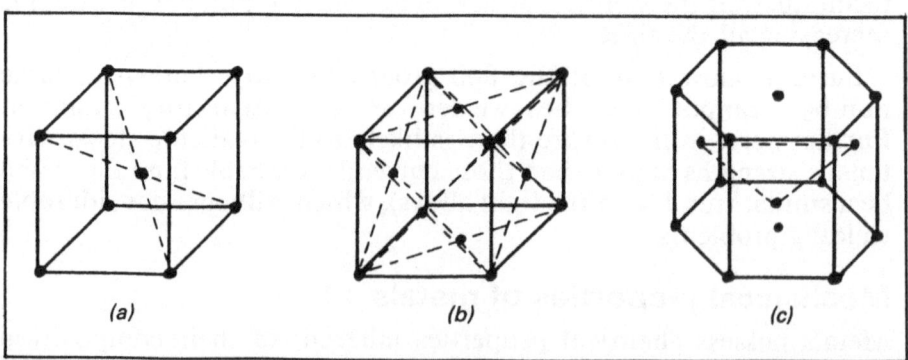

(a)   (b)   (c)

Fig. 4.1   *The most common crystal patterns: (a) body-centred cubic lattice, (b) face-centred cubic lattice, and (c) hexagonal close-packed lattice. The majority of solid metals have their atoms within these three systems*

whose shape and orientation are influenced by the environment in which solidification takes place, are called grains.

When metals solidify in large ingot moulds, the formation of grains can be observed in two characteristic arrangements: dendritic and columnar. At the edge of the mould where there is a strong directional heat flow, cooling is rapid and nucleation is plentiful with the result that sideways dendritic growth is restricted. This results in columnar type of growth which is commonly seen in solidified weld metal (Fig. 4.2a). In the centre of the mould where cooling is slow and the growth of the crystals unrestrained, true dendritic (fern-like) growth may be seen (Fig. 4.2b). The original dendritic structures, which may be responsible for some diminution of mechanical properties, can be modified by heat treatment. The atomic forces and the grain size influence the mechanical properties of metals.

Stress
When a force is applied to a metal, the latter is said to be under stress. The stress is the quotient of the force applied to the surface and the area of that surface:

$$\text{stress} = \frac{\text{load } (N)}{\text{area } (\text{mm}^2)}$$

Three main types of stress can be exerted on metals:

- *Shear* – tending to provoke slippage of one metal, or crystal plane in the metal, relative to another (Fig. 4.3a).

- *Compressive* – tending to force the atoms together (Fig. 4.3b).

- *Tensile* – the force tending to separate the atoms from each other (Fig. 4.3c).

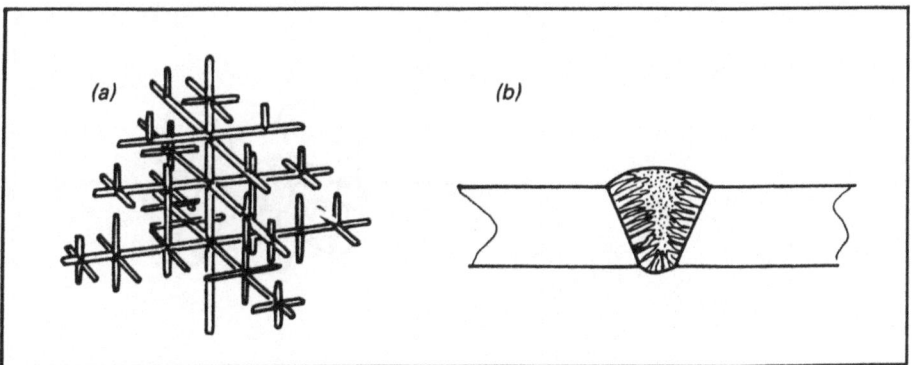

Fig. 4.2 Metal grain growth from the liquid : (a) dendritic and (b) columnar

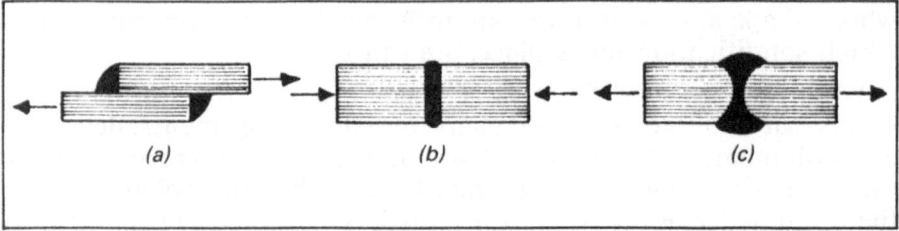

Fig. 4.3   The three main types of stress: (a) shear, (b) compressive and (c) tensile

Elastic and plastic deformation

As described earlier, the atoms within each metal grain are arranged in layers within a crystal lattice.   These layers have different orientations in different grains (Fig. 4.4a), and where the grains meet (the 'grain boundaries') the arrangement of atoms is disordered.   If the metal is subjected to a tensile stress this will act on the grains in different ways, depending on their orientation, with most grains being subject to some shear stresses (Fig. 4.4b).   Thus, whatever the external stress, the layers of atoms will have a tendency to slide over each other.

Under a relatively weak tensile stress, a metal distorts slightly as the atoms are forced to separate.   If the stress is removed, the interatomic forces return the atoms to their equilibrium positions and the metal will return to its original length (i.e. it has been subject to an 'elastic' distortion).

When the stress is higher and exceeds a certain limit, the atoms move relative to each other and the metal is no longer able to regain its initial shape (i.e. it has been subject to a permanent 'plastic' distortion).   The stress from which the metal will not return to its initial length is called the 'elastic limit'.

The atomic movement, during plastic deformation, is restrained from continual and easy movement by the atom disorder at the grain boundaries.   For this reason, metals become more resistant to

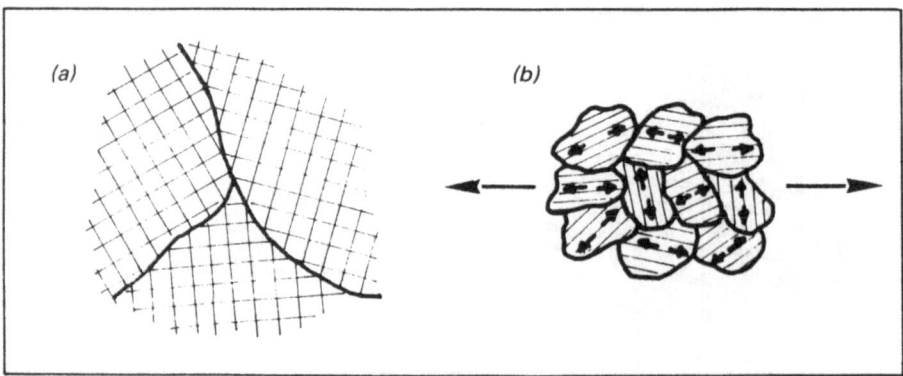

Fig. 4.4   (a) Grain boundaries and (b) under tensile stress

deformation, and increase in strength as the grain size becomes smaller. Larger grains, and lower strengths, may result from certain heating cycles where some grains can grow at the expense of others. The temperature at which this occurs varies from metal to metal. As far as welding is concerned, this is very important, particularly in the 'heat-affected zone' (HAZ) where the temperature is often sufficient to cause grain growth, and loss of properties; hence the importance of the choice of welding process and parameters because of their influence on on thermal cycles.

Tables 4.1 and 4.2 present some information of relevance to the topics discussed previously.

Table 4.1   Physical properties of steels

Fusion temperature
Pure iron: 1,538°C
Low alloy steels: 1,430–1,500°C
Austenitic stainless steels: 1,370–1,450°C

Thermal conductivity
(Measures the speed of heat propagation)
Pure iron: 81W/m/°K
Low alloy steels: 32–66W/m/°K
Austenitic stainless steels: 15W/m/°K
(These values are given at 20°C; thermal conductivity diminishes as the temperature increases)
Relative thermal conductivity (copper = 1): 0.17 for mild steels

Specific density
For all steels: 7,800–8,000kg/m³ (at 20°C)

Thermal expansion coefficient
(Measures expansion relative to an increase in temperature of 1°C)
Mild steels: 11.5×10^{-6}/°K (at 20°C)
Austenitic stainless steels: $14-15 \times 10^{-6}$/°K (at 20°C)

Specific heat
(Amount of heat to increase by 1°C the temperature of 1kg of metal)
490J/kg/°C (at 20°C)
(Specific heat increases with temperature)

Electrical resistivity
Mild steels: $10-20 \times 10^{-8}\Omega$m (at 20°C)
Austenitic stainless steels: $71-79 \times 10^{-8}\Omega$m (at 20°C)

Table 4.2   Melting points of selected metals and alloys

Aluminium	660°C
Bronze	900°C
Brass	940°C
Copper	1090°C
Steel	1430°C
Nickel	1450°C

Chapter Five

MECHANICAL TESTING OF WELDS

A N INVESTIGATION of the mechanical properties of a welded joint gives data which, if to specification, ensures the safety of the structure and, if poor, indicates welding faults and enables corrective action to be put in hand.

Mechanical testing of joints may be carried out either on a full-scale assembly, on test-pieces taken from the assembly (e.g. from run on tabs), or on test-pieces prepared under conditions identical to those of assembly manufacture.

Tests may also be performed using:

- Test-pieces consisting entirely of weld metal (produced as pads or in moulds). The properties determined will therefore be characteristic of undiluted weld metal.
- Test-pieces taken from welded assemblies where there will normally be some dilution of the weld metal by parent plate which is dissolved in the weld pool during welding. Dilution is affected by the operating conditions, such as arc current, arc voltage, electrode diameter, welding speed, welding position, preparation type, root gap, etc., and may vary from 0% (as in braze welding) to 100% (as in autogenous welding).

Tensile testing

The tensile strength of a material is a measure of its resistance to being pulled apart. This is determined by applying a steady load to test-pieces having dimensions defined by standards (e.g. BS709:1983). They are manufactured with ends, suitable for gripping in the chucks of the test machine, separated by the test length of reduced circular or prismatic section (Fig. 5.1a). A gauge length is marked on the test section to enable the elongation and reduction in area, occurring during the test, to be determined (Fig. 5.1b).

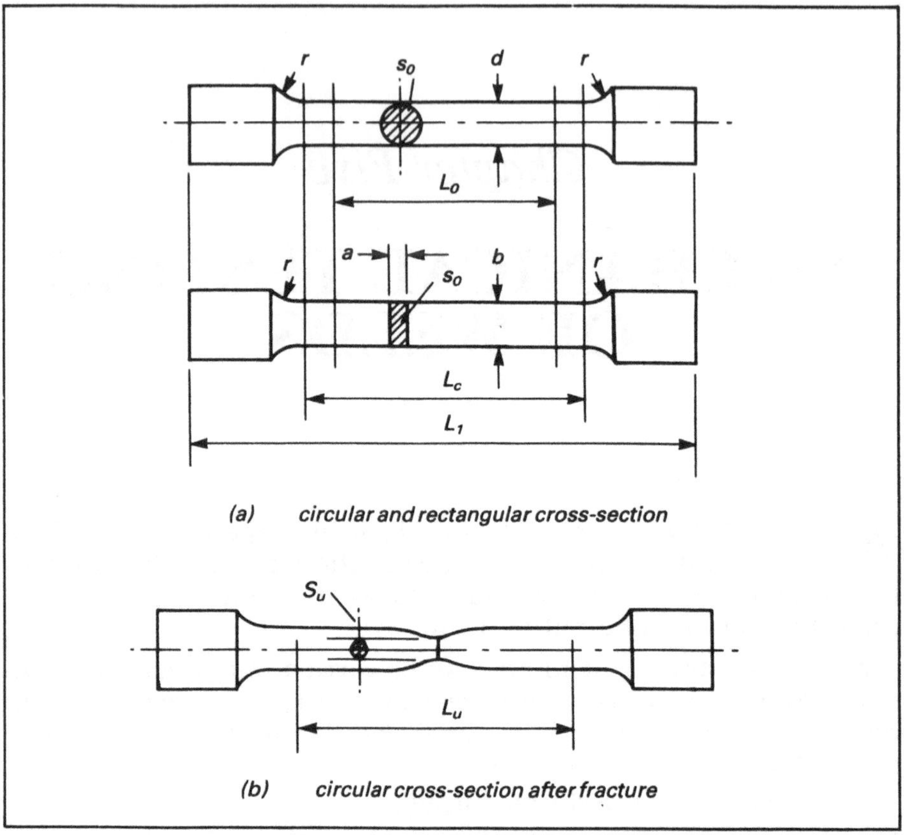

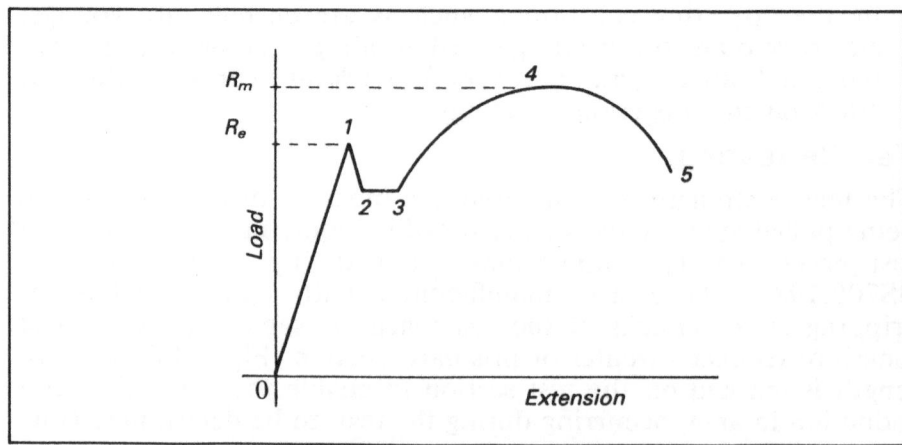

Fig. 5.1   Test-pieces of circular and rectangular cross-section

We can study the tensile test by first considering a idealised mild steel strain (load/deformation) curve (Fig. 5.2). This curve exhibits the following behaviour:

Fig. 5.2   Stress strain curve for a mild steel (idealised)

- From the origin to point 1 the curve is straight, indicating that the strain is proportional to the applied stress, i.e. point 1 is the elastic limit. In this example, point 1 is also the upper yield point (R_{eH}) – the point at which the material begins to deform plastically suffering permanent distortion. With many materials there is a region between the elastic limit and the yield point where the strain is not proportional to the applied stress.

- From point 1 to 2 the metal undergoes severe plastic distortion which can be maintained with a lower level of stress – the lower yield stress or yield point (R_{eL}).

- There is then often a level portion to the curve (point 2 to 3), where a large amount of strain occurs – yield stress elongation – for no increase in stress.

- From point 3 to point 4, the tensile strength (R_m) of the metal, the stress needed to produce plastic strain continues to increase. At point 4 the extension continues but the cross-section of the test-piece begins to decrease, usually locally, and is said to 'necking'.

- Thus, beyond point 4 the stress per unit area can increase, even though the actual stress can no longer be supported and, at point 5, it fractures.

Most metals and alloys behave in a manner different from mild steel when subjected to a tensile test. Not only is the stress strain curve continuous (Fig. 5.3a) but also for a number of materials, particularly the aluminium alloys, there is no straight portion of the curve (Fig. 5.3b), i.e. the strain is not linearly proportional to stress. For these materials the 'proof stress', and not yield stress, is the factor used in design   calculations.

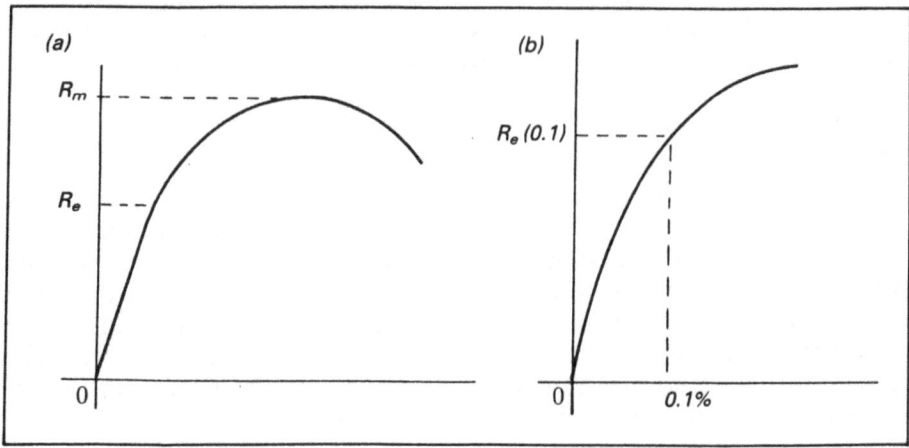

Fig. 5.3   (a) Stress strain curve typical of many metals and alloys and (b) stress strain curve for aluminium alloys

Proof stress is defined as the stress at which a non-proportional elongation, equal to a specified percentage of the original gauge length, occurs (Fig. 5.3b). When a proof stress (R_p) is specified, the associated non-proportional elongation should be stated (e.g. 0.2%).

A measure of the ductility of the material is the elongation of the test-piece at fracture. It is recorded as 'percentage elongation' as defined below:

$$\text{Percent elongation} = \frac{\text{increase in gauge length}}{\text{original gauge length}} \times 100$$

The interaction of the weld metal, the heat-affected zone and the parent metal in welded specimens affects the values of elongation and elasticity so that they have limited absolute value. As a result, such tests are usually performed on 'all weld metal' and on parent metal test-pieces. Welded joints are tested to evaluate their compliance to a standard (e.g. BS4515) and, in meeting the standard, weld integrity is proven. (A 'joint coefficient' might be defined as the ratio between the breaking load of a welded test-piece and a similar test-piece of parent metal.)

A second measure of the ductility of the material is the reduction in cross-sectional area which occurs during the tensile test, and is defined as follows:

$$\text{Percentage reduction in area} = \frac{S_o - S_u}{S_o} \times 100$$

where S_o is the original cross-sectional area of gauge length and S_u is the minimum cross-sectional area after fracture.

The ductility of metals as measured by elongation, ranges from low at, say 5% to good at 30%. Materials with little or no ductility are said to be brittle. Glass, an example of such a material, is slightly elastic, as we can observe in the deflection obtained when light pressure is applied to a window pane, but it breaks in a brittle manner when the load is increased.

When a ductile material fractures, the surface of the break is typically irregular and torn and has a fibrous appearance. The fracture can be extremely slow, taking seconds, hours or even months to spread across the whole metal section. Brittle fracture on the other hand occurs very rapidly, in fractions of a second. The material will often fracture into several pieces with the fracture generally clear and bright and usually smooth although crystalline in appearance (but often with shredded edges).

Metals become more brittle as the temperature decreases, because of increasing rigidity of the crystalline networks which resists relative movement of the crystal planes. This change is not abrupt but occurs progressively over a range of temperature known as the 'transition range'.

In general, as tensile strength is increased, say by working, ductility falls.

Fracture toughness

A tough material is able to absorb large amounts of energy without breaking. It is to some extent a property resulting from a combination of tensile strength and ductility. The area under the stress strain curve is some indication of the toughness of a material. However, as the load is only applied slowly during a tensile test, this measure is only a qualitative indication of the ability of the material to deform under rapid or impact loading. Indeed some materials, exhibiting good ductility in tensile tests, fail in a brittle manner when stress is applied at high rates as they are unable to absorb the received energy.

There is a need to determine some relative measure of material toughness as impact loading is often found in real life. This test must also consider the affects of material 'defects', such as notches, cracks and sharp changes of section, as they have a considerable effect on behaviour inder impact. As a result, the test-pieces developed for toughness tests have an artificial notch of U- or V-shape. Perhaps the most universal is the Charpy-V test, with the V-notch being chosen because of its relative severity.

In principle, the energy absorbed is determined by a single blow from a pendulum swinging onto a notched specimen (Fig. 5.4). Tests

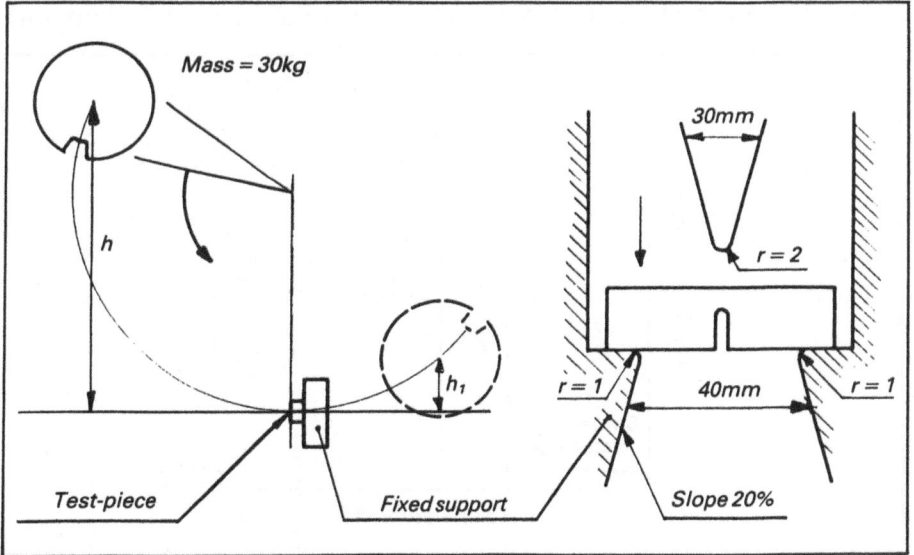

Fig. 5.4 Swinging pendulum Charpy impact test

are carried out at specified temperatures and the standard specimens are notched at mid-length. The height of the pendulum before the test determines the applied load/energy. The falling pendulum strikes and fractures the specimen and the rise on the follow through indicates the residual energy. Simple subtraction gives the energy absorption, usually expressed in Joules. The test report must also include information on the dimensions of the test specimen, the location and orientation of the notch and the test temperature, and give a description of the fracture surfaces. Details of test-piece dimensions and the test procedure are given in BS131, Part 2 and BS709:1983.)

By careful placement of the notch, the various zones of a weld (e.g. weld metal, fusion boundary or HAZ) can be tested. A metal or regions of a metal where toughness decreases abruptly when it includes a notch is said to be 'notch sensitive'.

Materials become more brittle and lose toughness as the test temperature is reduced. The temperature at which material fracture changes from tough to brittle is known as the 'transition temperature' (Fig. 5.5).

Hardness testing

Hardness is the property of a material to resist penetration. As a test technique, a pyramid, conic or spherically shaped indentor is forced into the material under a given load, an indentation is produced and the hardness is taken to be inversely proportional to the amount of indentation. It is perhaps the most common test made because of its simplicity and because it is relatively non-destructive. Also, for steels, there is a close relationship between hardness and yield and

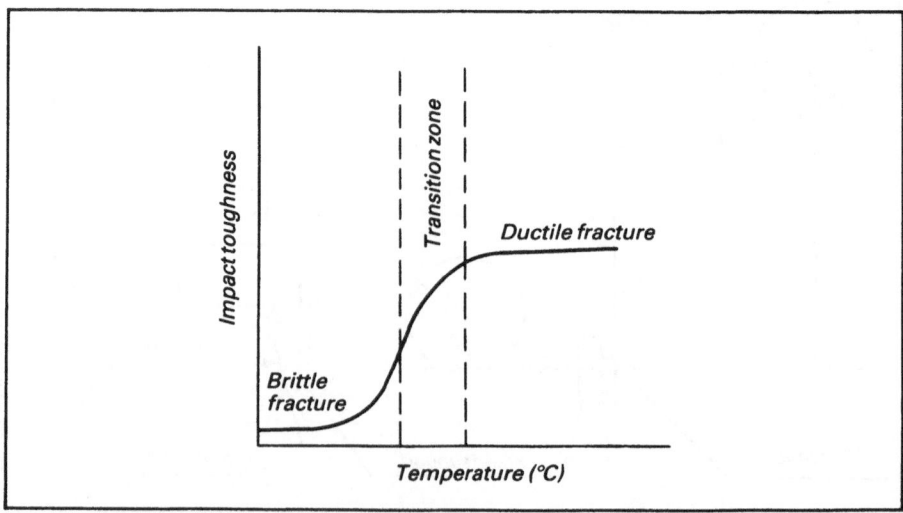

Fig. 5.5 Impact toughness transition curve

tensile strengths. The three best known hardness tests are the Brinell, the Rockwell and the Vickers.

In the Brinell test a hardened steel, or tungsten carbide, ball of 10mm diameter is forced into the surface of the metal under a load of 3,000kg (or 500kg for softer metals). The diameter of the imprint left by the ball is measured, the surface area calculated and the Brinell hardness number (BHN) determined:

$$\text{Brinell hardness number (BHN)} = \frac{\text{load}}{\text{area of indentation (mm)}}$$

The Brinell machine produces a relatively large indentation (with a large load) which makes it unsuitable for measuring the hardness of thin metals, plated surfaces, surfaced-hardened materials, etc. However its use is desirable when the average hardness is required of non-homogeneous materials, e.g. cast iron.

Rockwell hardness is determined from depth of indention measurements. A small steel ball or a conical diamond penetrator is used. In the test, a 'minor load' of 10kg is first applied to the indentor, the depth of a penetration dial gauge set to zero, and then a major load of 60, 100 or 150kg added. The major load is then released and the dial gauge reading the Rockwell C hardness noted.

There are a number of standard Rockwell hardness test ranges, based on specified minor and major loads, which give this technique great versatility.

The Vickers test is made with even lower loads, typically 5–120kg, and uses a square-based pyramid-shaped diamond indentor. As with the Brinell test the hardness is the quotient of the load and the surface area of the indent. The Vickers pyramid number (VPN) is usually obtained from tables after the indentation diagonals have been measured.

The Vickers test may also be performed with a wide range of loads and indentation sizes and is particularly suitable for measuring small areas (as a micro Vickers test even microstructural phases), and is widely used to measure weld hardness.

The hardness of a weldment is usually determined on macro or micro cross-sections. A number of hardness indentations are taken in defined (e.g. BS709) regions including parent metal, HAZ and weld metal (Fig. 5.6).

Weldment hardness can be discussed with reference to Fig. 5.7. This weld is made in low alloy steel (carbon (0.15%), chrome (0.40%), nickel (2.30%) and molybdenum (0.20%)) of 45mm thickness. Following deposition of a weld bead, the material was transverse-sectioned, polished and a traverse of Vickers hardness tests performed. When welded without preheat, the hardness varies from

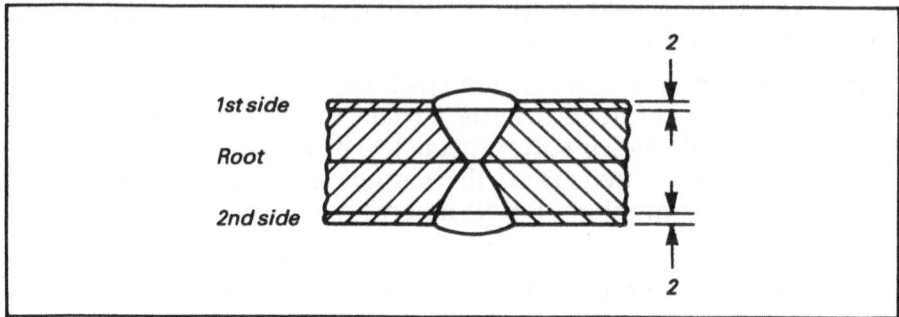

Fig. 5.6   Hardness test on a weld joint (after BS709:1983)

170 in the parent metal to 280 in the HAZ (curve A, Fig. 5.7). This HAZ hardness was unacceptable for the application and a preheat of 150–200°C was applied before repeating the test. This had the effect of reducing the HAZ hardness to below 200 (curve B, Fig. 5.7). (The effects of preheat are discussed further in Chapter Seven.)

Bend tests
In this test, a section taken from a butt welded joint is bent around a circular former, enabling the soundness of the weld metal, weld junction and HAZs to be assessed. The test also gives some measure of the ductility of the weld zone and may be made transverse to or along the longitudinal length of the weld joint. The test may be performed either by rolling the specimen around a former of specified diameter or by forcing around a former as shown in Fig. 5.8.

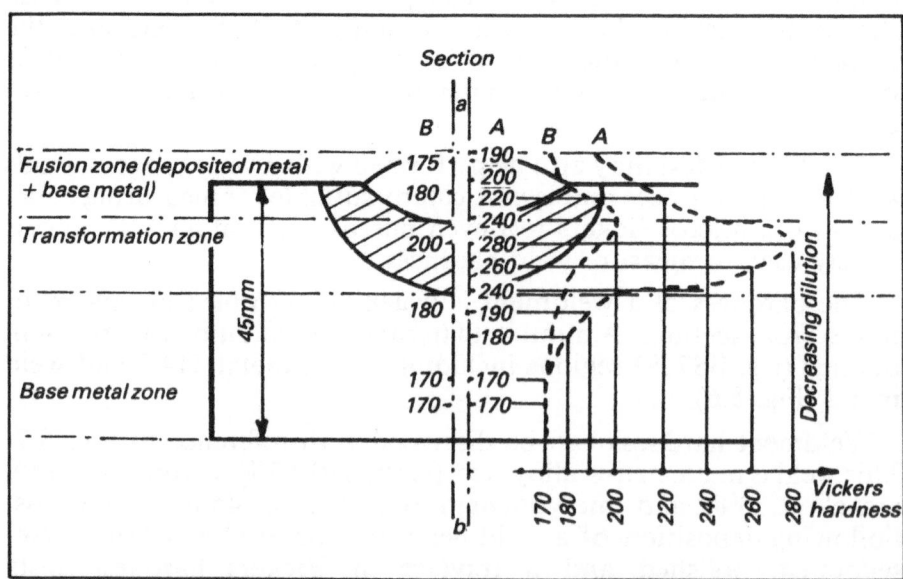

Fig. 5.7   Results of Vickers hardness testing of weld joint

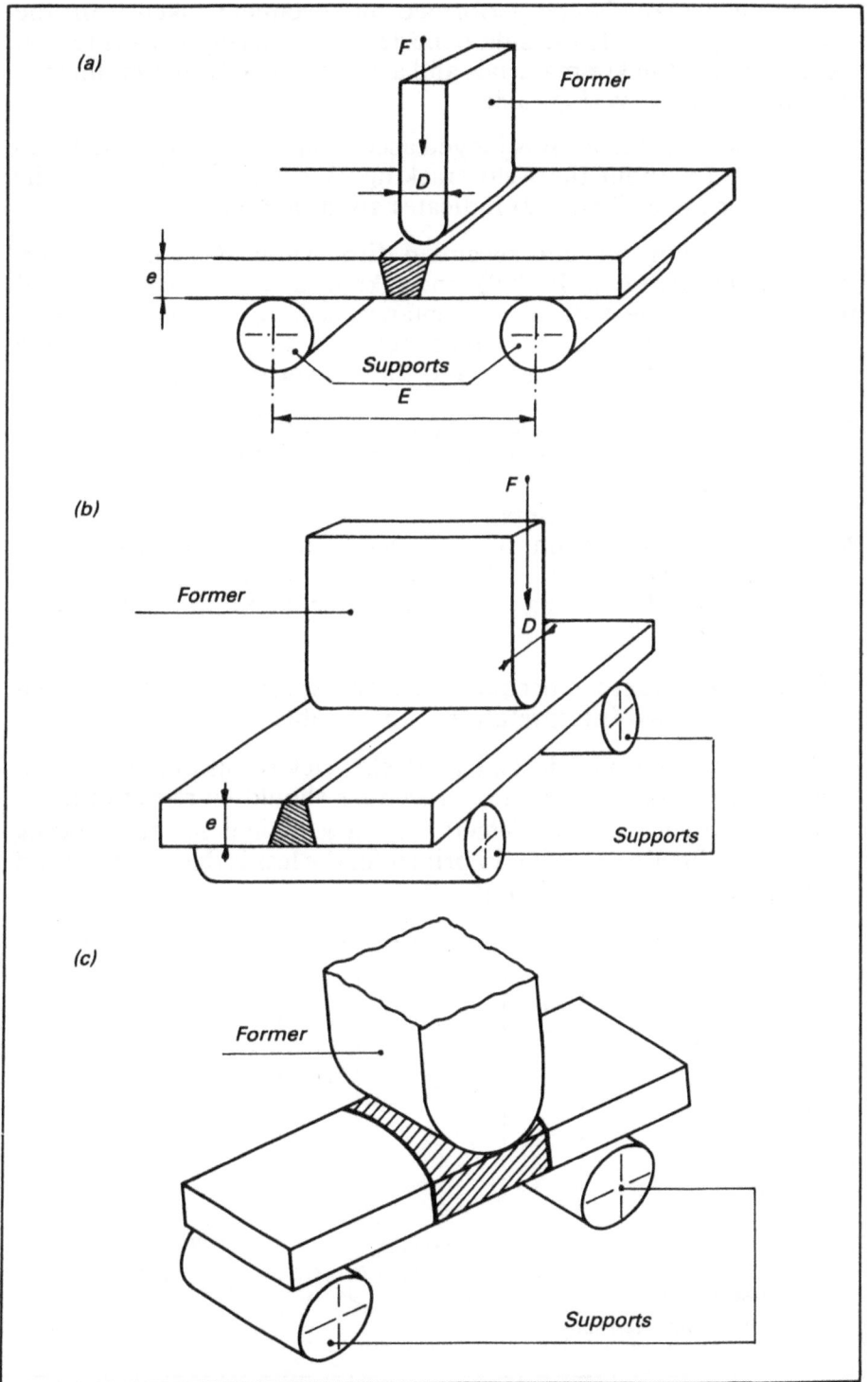

Fig. 5.8 Bending tests: (a) transverse, (b) longitudinal and (c) side

Bend tests may also be performed on specimens taken from the weld cross-section. These side bend tests are usually performed on welds made in thicker materials and are done in a similar manner to the other bend tests (Fig. 5.8c).

The quality of the joint is judged according to the bend which can be achieved without the weld cracking. A bend of 180° where the limbs are parallel (Fig. 5.9) indicates good ductility.

The test conditions and specimen dimensions and condition are fixed by standard (e.g. BS709), and take into account the nature of the metal being tested. The acceptance criteria is either specified within the standard or is agreed between the customer and the manufacturer. The test results and their severity depend upon:

- The former dimension being equal to $1e$, $2e$, $3e$, $4e$, etc. (e = plate or sheet thickness) with the severity of the test decreasing progressively.
- The spacing of the rollers.
- The surface condition of the specimen (e.g. rounded edges, polished surface, etc.) – which is very important.
- The speed of bending – a parameter often not sufficiently considered.

If specimen fracture occurs before the specified bend has been achieved then two interpretations are possible:

- The joint contains defects, e.g. cracks, lack of fusion, inclusions, etc. In this case the welding procedure should be re-examined.
- In the absence of any defects, the lack of appreciable bending indicates brittle material. In principle, the less is the angle of bend the greater is the brittleness.

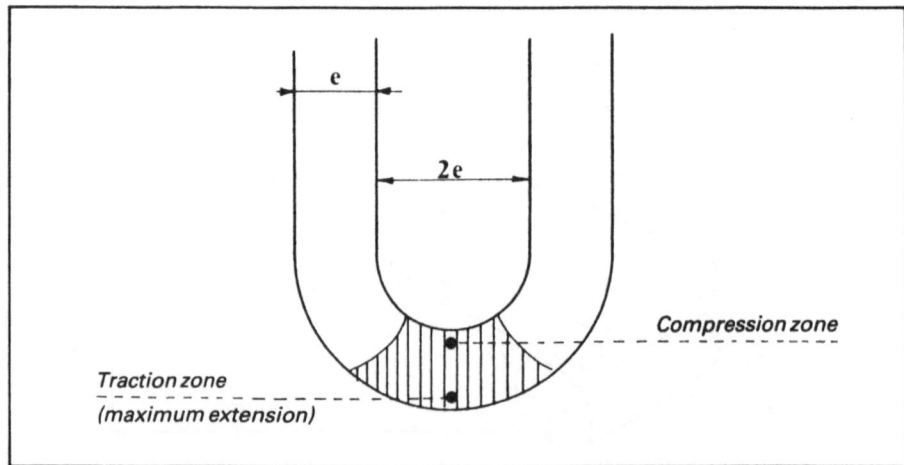

Fig. 5.9   Weld face bend: of 180° using a former of diameter 2e

The appearance of the fracture surface may also be significant, although careful interpretation is required:

- A bright appearance often corresponds with coarse grain size and or brittle failure.
- A dull appearance with a good fine-grained structure.

Nick break test

This economical test (Fig. 5.10) shows up the defects of a weld. It consists of:

- Taking a section from the weld.
- Making a small notch (to initiate fracture) in the axis of the weld.
- Fracturing the test-piece by bending or hammer blows.
- Examining the fracture face.

The test report will cover the condition and texture of the fracture face and will note the presence of defects, such as cracks, lack of fusion, lack of penetration and inclusions.

Peel and torsion tests

These tests are in general used to destructively test resistance spot welds and will not be considered here.

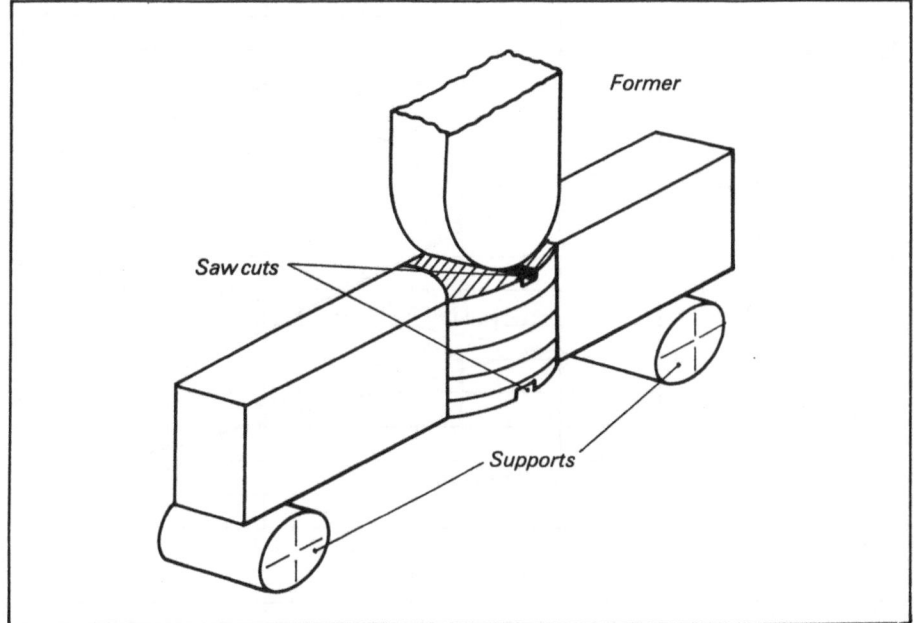

Fig. 5.10 Nick-break test

Relationships between tensile strength, yield strength, elongation toughness and hardness

For each type of material there exists relationships between these properties, although not always quantitative. Tensile strength, elasticity (yield strength) and hardness usually vary in the same direction. For example, in steels, they increase with increasing carbon content (Fig. 5.11). The elongation and toughness also tend to vary in the same, but opposite, direction to the other properties.

Metallographic examination of metals

The science of metallography is essentially the study of the structural characteristics or constitution of a metal or an alloy in relation to its physical or mechanical properties. Such a study or examination of metal may be macroscopic or microscopic. Macroscopic examination involves visual observation of the gross structural details of the material, either by the unaided eye or with the help of low-power magnification, such as a magnifying glass, binocular or low-power (generally less than ×10) microscope. Microscopic examination, usually requiring the specimen surface to be prepared, employs optical or other forms of microscope to permit examination to be made at higher magnifications. The observed structures may, in both instances, be photographed.

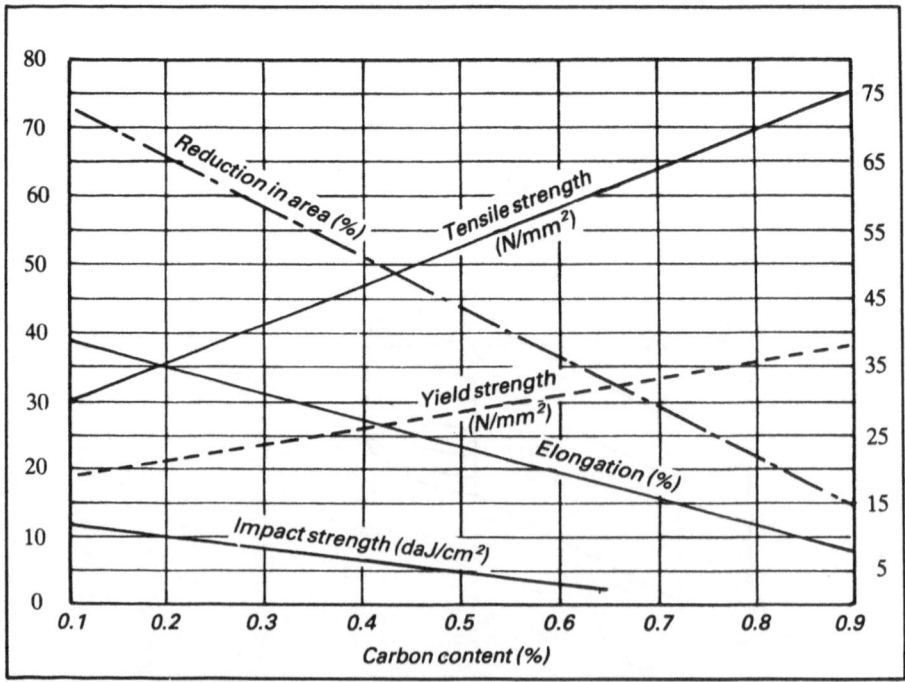

Fig. 5.11   Effect of carbon content on the mechanical characteristics of a hardened steel

Macroscopic examination

Macroscopic examination can be used to observe the gross structure, e.g. the penetration profile, of welds. To do this, it will be necessary to section the weld. This is usually performed by sawing, as heat from, for example, flame cutting can modify the structure. The cut surface is then ground and polished using fine abrasive papers. Next, the test-piece is cleaned and etched to show the different weld zones. Etchants can be aqueous solutions of acids or mixtures of acids, e.g. 10–20% nitric acid for steels and 25% for non-ferrous metals*. Macro examination will also show major defects, such as lack of fusion, inclusions and cracks.

Microscopic examination

For microscopic examination, the surface preparation is commenced as for macroscopic examination but must then be taken through several further stages. First finer and finer grades of abrasive paper are used to polish the metal surface. The surface is then polished to a mirror-like finish using various grades of diamond powders. The etchant used is carefully chosen to selectively attack or highlight the appropriate parts of the microstructure. In general, etching has the following effects:

- The grain boundaries are more strongly attacked than the grains themselves and the latter are therefore put into relief.
- Those phases or constituents of the microstructure attacked least reflect light better and thus appear brighter.
- Within the same constituent, the degree of attack and hence the contrast varies with the orientation of the grains.

These combined effects allow the micro constituents to be identified, which in turn gives information about the heating cycle it has experienced, about the alloying elements it contains, the material properties (hardness, strength, toughness, etc.) and its likely corrosion resistance. Microscopic examination also allows the detection of defects such as cracks and fine inclusions.

Corrosion behaviour

Corrosion is the alteration of a metal as a result of the chemical combination of one or more of its components with other elements. Corrosion may occur in gaseous or liquid environments, with air, especially when the humidity is high, being a prime example. Thus oxidation or rusting, which is a form of corrosion, is present in all metallic constructions. The forms of protection used to prevent or inhibit such oxidation have importance for the welder, either because the coatings affect the process behaviour or because the parent material composition is altered, as with the use of stainless steels.

---

* In laboratory testing, 10% nitric acid in industrial alcohol is a common reagent. At all times care must be taken in mixing acid solutions especially when acid concentration is greater than 10%.

Corrosive attack takes many forms. It may be general or localised to specific alloy constituents, to the grain boundary regions (intergranular corrosion), or within cracks or crevices. Corrosion is generally aggravated by stress (stress corrosion) which is always induced by welding.

Steel products are often coated with paint or primer to protect them from rusting and this is often performed before welding, possibly at the rolling mills. Ideally it should be possible to weld over such primers without affecting weld quality. At other times, it will be necessary to clean the weld preparation area prior to welding to avoid welding process problems such as porosity.

The oxide of aluminium, which forms very quickly even at room temperature, differs from that of steel in that it adheres very strongly to the underlying metal, and once formed limits further oxidation. To produce sound welds in aluminium and its alloys it becomes necessary to use techniques which will disrupt this oxide film.

The use of nickel and chromium as alloying elements in metals developed to resist oxidation at high temperatures, requires modifications to the welding processes used, e.g. in terms of the filler material or shielding gas composition.

Chapter Six

WELDABILITY

THE ORE from which iron (Fe) is extracted is a collection of oxides (FeO, Fe_3O_4, Fe_2O_3) which are reduced (i.e. the oxygen is extracted) in a blast furnace by passing through the ore a current of carbon monoxide (CO) which is produced from the reaction of carbon (coke) and oxygen (air). The product is iron and carbon dioxide (CO_2) as a result of reactions of the form:

$$FeO + CO \rightarrow Fe + CO_2$$

However the ore contains elements other than iron, such as sulphur (S), phosphorus (P), manganese (Mn) and silicon (Si), which are also reduced and dissolved, together with considerable quantities of carbon, in the molten product from the blast furnace. This crude iron (similar to cast iron) is brittle and difficult to work. The metal is therefore purified to reduce the levels of these embrittling elements. This is done by passing a stream of oxygen over or through the molten crude iron to 'burn out' the impurities. The resulting oxides are absorbed into a slag which floats on the metal surface. The product which still contains residual elements and a controlled level of carbon is known as 'steel'. During this refining process, oxygen is also dissolved in the metal and if it is not removed before casting it will result in porous products. Its removal is achieved by adding back into the steel elements (deoxidants) which preferentially combine with oxygen, such as silicon and aluminium. When steel was produced in open hearth furnaces, the refining process resulted in the metal 'boiling' and when this was stopped through the addition of the deoxidants the boil was said to have been 'killed'. Hence the terms semi-killed and killed steels which refer to the levels of deoxidation carried out.

Steels – Metallurgical considerations

The molten steel is either continuously cast into bar, rod or plate form, or cast into moulds. When semi-killed steel is cast into moulds and allowed to solidify, there is a skin of pure porous-free metal formed against the mould walls while the core of the mould exhibits porosity. This steel (rimming steel) can be rolled into sheet which retains the two zones. It has good forming properties and has been widely used where metal sheet must be pressed into shape. During the solidification process the impurities, such as slag, are pushed in front of the liquid/metal interface and they therefore collect in the last portion to freeze the upper portion of the ingot. This region is usually cropped off before the next operation, e.g. rolling.

Carbon is an essential constituent of steel – its presence adding strength to the alloy. The iron carbon alloys may be divided into three categories:

- Irons in which the carbon content is very low and has a negligible effect on properties.
- Steels in which the carbon content is important and is usually in the range 0.1–1.5%.
- Cast irons in which the carbon content is such as to cause some liquid of eutectic composition (see later) to solidify. The minimum carbon content is about 2.0% while the maximum is about 4.5%

To explain some of the changes which occur in steels during welding, it is necessary to consider in outline the metallurgy of the iron/carbon system. Alloys composed of two elements may solidify according to one of the processes below:

- The elements possess different crystalline form and solidify independently, forming simple mechanical mixtures, e.g. lead and copper. There is no solubility of one in the other.
- The elements have a similar crystal structure and they can dissolve one within the other to form a 'solid solution'. There are two possibilities. If the atoms of the added element are much smaller than the atoms of the major element then they are able to distribute themselves in the inter-atomic spaces, i.e interstitial alloying. Small amounts of carbon and, as mentioned later, hydrogen, can interstitially alloy with iron. If the added atoms are about the same size (and of similar chemical behaviour) as the atoms of the host element, then they may directly replace certain atoms, i.e. they directly substitute.
- The elements may react or combine together to form an alloy or phase which will crystallise in its own crystal system.

Essentially, apart from the impurity elements which it contains, steel can be considered as a binary alloy of iron and carbon. Its solidification behaviour could be compared with the solidification and freezing of a strong salt solution. As we heat a salt solution we find we can dissolve more salt, i.e. the solubility increases with temperature. Now, as we cool and then freeze the solution, salt is first precipitated and is then trapped in a frozen water/salt alloy. This is roughly the same for carbon in (cast) iron, where the carbon is deposited as graphite within an iron-carbon alloy.

The behaviour is more complicated because iron, and hence steel, is an allotropic metal, i.e. its internal structure changes with temperature (Fig. 6.1). Pure iron solidifies at about 1,540°C into a body-centred cubic structure known as δ (delta) iron. On slow cooling it transforms in the solid at about 1400°C, to γ (gamma) iron which has a face-centred cubic structure. Yet another transformation occurs at about 910°C when α (alpha) iron is formed again with a body-centred cubic structure. The reverse of these transformations occur on heating and each is accompanied by an absorption (or, if cooling, a release) of heat which is, however, small when compared with that occurring at the melting/freezing point. The temperature at which the transformation occurs can be affected by the cooling rate. Furthermore, almost all the properties of the metal change when a metal transforms from one allotropic form to another.

The presence of carbon, or other alloying elements, in the iron alters the equilibrium temperature at which these transforms occur, as exampled for 0.20% carbon steel in Fig. 6.1. Carbon in gamma-iron gives a structure called 'austenite' and in alpha-iron one called

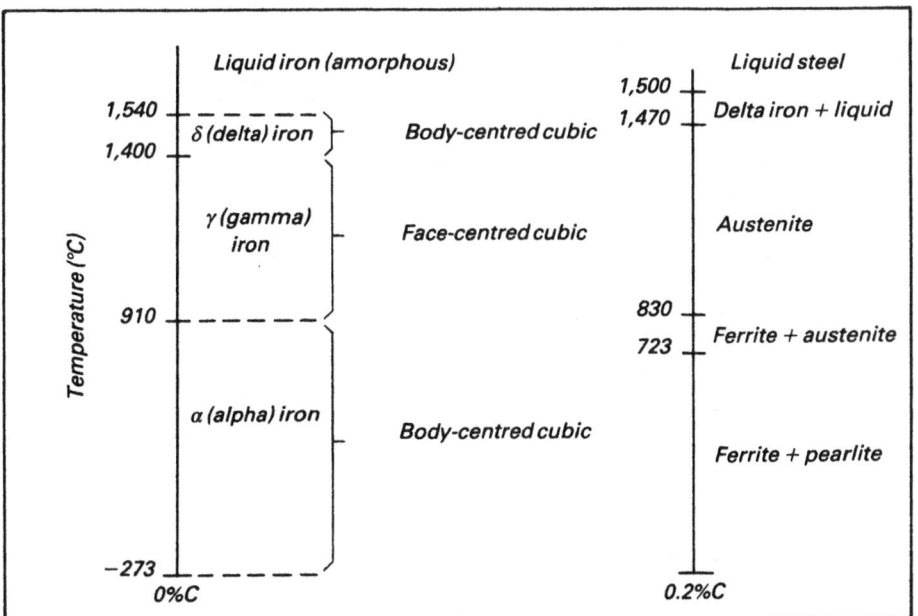

Fig. 6.1 The temperature ranges in which the allotropic forms of pure iron and iron containing 0.2% carbon exist under equilibrium conditions

'ferrite'. Let us consider the behaviour of such a 0.20% carbon steel during solidification and cooling. This is best done with reference to an iron-carbon equilibrium diagram, which in terms of temperature and carbon content describes the boundaries of phase changes (Fig. 6.2).

At about 1500°C (point 1) the steel starts to solidify; the line defining this transition is known as the 'liquidus'. As the temperature falls, atoms accumulate around nucleating groups to form islands of delta-iron. At about 1470°C there is no residual liquid and the delta-iron solid begins to transform to austenite (point 2). For most practical purposes the transformation through the delta phase can be ignored. As the metal cools, it remains austenitic until it crosses the A_3 line – the line defining the onset of the austenite to ferrite transformation. As the temperature is continuously reduced at a slow rate, more ferrite forms until the A_1 temperature is reached. At this temperature, the remaining austenite decomposes into pearlite and no noticeable changes occur on further cooling. Pearlite is a lamellar structure formed of alternating plates of cementite and ferrite, and as a mechanical mixture of two phases it has properties intermediate to

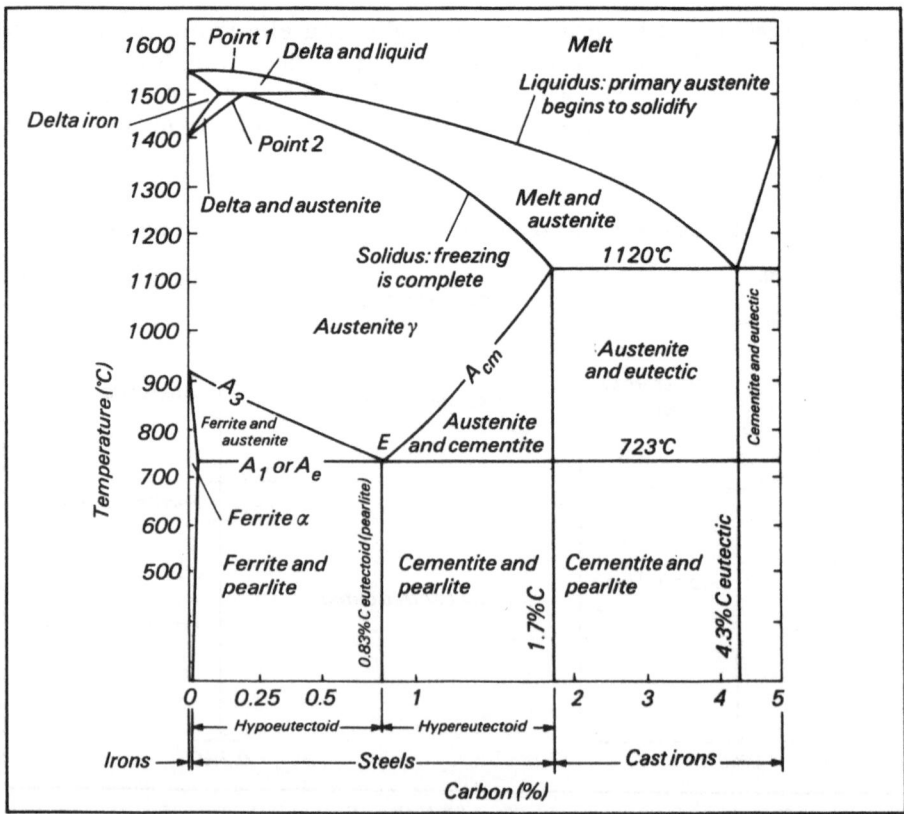

Fig. 6.2   Iron-carbon equilibrium diagram (adapted from 'Metals Handbook')

those of its components. It forms because ferrite has a low solubility for carbon (max. 0.05% at 723°C) compared with that of austenite (max. 1.7% at 1130°C). Thus as ferrite is formed, the excess carbon is absorbed by the deposition of cementite (6.7% carbon). As this formation takes place in the solid, the carbon must diffuse from ferrite to cementite and this is easiest with the platelet form of growth. Under microscopic examination, pearlite has a characteristic form not unlike the edge of a mother-of-pearl shell, hence the name pearlite.

When the heating of steel is considered, as during heat treatment or welding, the transformation commences at the A_1 line but is not complete until the A_3 line is reached, and thus is usually considered as the transformation or critical point. The interval between A_1 and A_3 is known as the transformation or critical interval.

The sequence described above only applies in detail to a steel containing 0.02% carbon. With other carbon contents the phases will still occur but at different temperatures. For example, for a steel with about 0.83% carbon, the final structure will only contain pearlite. The diagram also indicates that the liquidus-solidus interval increases for carbon contents up to 1.7% and decreases thereafter until the eutectic point is reached (4.3% carbon). This is the point at which the iron-carbon alloy passes from one state to another, on one side the primary solidifying grains being austenite and on the other cementite. The solidus line stabilises at 1,120°C for alloys containing between 1.7% carbon and 6.7% carbon. An alloy of 6.7% carbon is pure cementite. Steels with carbon contents less than 0.83% (eutectic point) are said to be hypoeutectoid while those with carbon levels up to 1.7% are hypereutectoid.

It is essential to note that the transformations described occur under very slow (equilibrium) cooling conditions. If solidification and cooling is very rapid, which is generally the case in welding, these transforms may occur with a delay. Indeed, with the addition of certain alloying elements, they may be retarded, suppressed or made more active.

It is also interesting to note that the melting temperature at the eutectic point of an alloy is lower than the melting point of either of the primary constituent elements. Thus, for example, in the field of non-ferrous metals, the melting point of the eutectic composition of the tin-lead alloy (63% Sn, 37% Pb) is 183°C compared with 231°C for tin (Sn) and 327%°C for lead (Pb).

Influence of alloying elements and impurities in steel
Steels are iron-base alloys to which the addition of specific elements and the presence of impurities produce particular and precise mechanical and chemical properties.

Alloying elements are those added deliberately for a specific reason. Impurities are those which are undesirable and can arise from the air (oxygen and nitrogen), the base metal (sulphur and phosphorus), or from the atmosphere surrounding the metal during its preparation (hydrogen coming from the breakdown of moisture).

Impurities found in steels are absorbed in varying degrees into the grains during solidification. Those not absorbed find their way into the grain boundaries where they generally have a detrimental effect on mechanical properties. They are often from low melting point constituents which lead to defects when stressed hot, i.e. hot cracking. Steels are more sensitive to hot cracking, the higher their carbon content and impurity level.

As noted previously, sulphur and phosphorus are two of the principal impurities found in iron. Sulphur if combined with iron forms a brittle low melting point constituent which deposits on the grain boundaries and is very detrimental to the transverse strength and impact resistance. It also adversely affects weldability. However, sulphur combines preferentially with manganese to form stringer inclusions which cause little harm (they can improve machinability). Even so, steelmakers usually keep sulphur below 0.05% (except where sulphur is added to improve machinability).

Phosphorus is also kept below 0.05%. If the content becomes too high, the iron phosphide formed causes intergranular embrittlement severely decreasing ductility and toughness.

There is a trend in modern steelmaking to restrict both the sulphur and phosphorus to lower levels (e.g. 0.02%). During welding, the effects of such impurities, particularly when dilution is high, is accentuated by the high cooling rates and the induced stresses.

Dilution is defined as $A/(A + B)$, where, on a weld cross-section, A is the area of deposed metal and B is the area of parent metal fused.

The presence of hydrogen, particularly in weld metals, can lead to serious problems. Hydrogen is an extremely mobile element in steel and other metals. It can arise from the breakdown of moisture found in or on fluxes, wires or parent metal surface, and from the natural or protective atmosphere surrounding the weld. During solidification, the relatively small atoms of hydrogen readily insert themselves into the face-centred cubic austenite structure. The solubility of hydrogen in molten steel is great, and when the metal freezes rapidly the solid can easily become saturated with the element, in extreme cases causing porosity. In the solid, the solubility is greater in the austenite phase than in the ferrite (Fig. 6.3) and therefore even though hydrogen can rapidly diffuse through the solid (especially when hot) these two decreases in solubility can lead, at ambient temperatures, to significant and detrimental hydrogen build up.

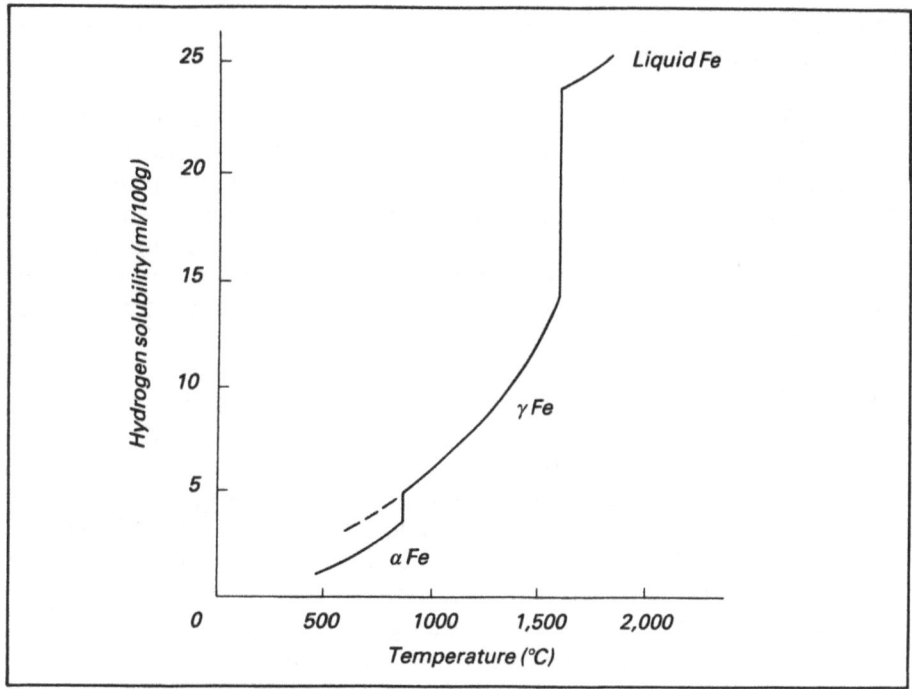

Fig. 6.3 Solubility of hydrogen in weld metal

Hydrogen causes embrittlement of the structure and by diffusing into discontinuities can develop enough pressure to form hairline cracks in the steel (hydrogen embrittlement and hydrogen cracking). Both these effects are even more severe in the presence of martensitic structures (formed if austenite is rapidly cooled preventing the transformation to ferrite and pearlite). As the pressure build up due to diffusion takes time, the cracks often develop some time after the high temperature event, e.g welding, and the phenomenon has therefore also been called cold cracking, delayed cracking or, in welds, underbead cracking.

Additive and alloy elements
Simple carbon (mild) steels, containing less than 0.20–0.25% carbon, pose relatively few welding problems. However, it becomes necessary to adopt various precautionary measures (e.g. pre-heat) as the hardenability of the steel increases. Hardenability should not be confused with the hardness as such or with maximum hardness. Hardenability is the property that determines the depth and distribution of hardness induced by quenching and could be said to be related to the ease with which a martensitic structure is formed. The more hardenable a microstructure is the more sensitive it will be to cracking, and the lower will be its weldability.

Hardenability and weldability will be affected by those elements which are added, at low levels to low carbon steels, to improve mechanical properties. It may be useful to outline briefly the role of the principal elements.

Carbon (C). This is the most influential element and has a dual effect on hardenability: it controls the maximum attainable hardness and contributes to hardenability. It is one of the more potent and economical strengthening elements. Increasing amounts of carbon, together with the presence of other elements, such as manganese, chromium and molybdenum, promote the formation of martensite. The higher the carbon content the harder will be any martensite that is formed.

Manganese (Mn). This is a very important element because of its strengthening effects and its ability to combine with and minimise the effects of sulphur. In carbon steels its content varies from 0.3–0.8% and in carbon-manganese steels having higher yield strengths may attain 1.7%. At high levels of 12–14%, manganese inhibits, when the steel is quenched, the formation of martensite which subsequently forms when the steel is worked. This steel is very wear resistant. The addition of manganese produces a fine grain structure which has greater toughness and ductility.

Chromium (Cr). Chromium is an active hardening element which increases toughness in low alloy steels. When added alone (i.e. additionally to carbon and manganese which are always present), its content is usually below 1.0%, at least in those steels said to be readily weldable. At higher contents, especially as the carbon content increases, the hardenability is such that precautions are required if welding is to be successful (e.g. preheating must be applied). An important group of chromium-containing alloys, developed for their resistance to temperatures in the range 450–500°C, also contain molybdenum (nominally less than 0.2%C, 0.5–10%Cr and 1–1.25%Mo). For welding, preheating is required. At high levels the properties are intimately linked to the carbon content. With the higher carbon contents, chromium carbides are formed which are highly abrasive. These alloys pose considerable welding problems. At lower carbon contents (0.15%) and with a nominal 13%Cr, martensitic stainless steels are produced. Weldability is very much related to carbon level. For instance, the low level alloys, as used for their heat resistance by the automotive industry, are readily welded. The combination of high chromium plus nickel (e.g. 18%Cr, 8%Ni) gives the austenitic range of stainless steel alloys which are generally weldable.

Nickel (Ni). Nickel is used in low alloy steel to strengthen and improve low-temperature toughness and to increase hardenability. It also appears to reduce the sensitivity of the steel to variations in thermal cycle which is generally beneficial. It is particularly effective when used in combination with chromium and molybdenum. It is used in air-hardening steels which have low weldability, in steels used at very low temperatures (i.e −190°C) and in rust-resisting steels.

Molybdenum (Mo). This is another active hardening element which is useful in maintaining the hardenability between specified limits. In amounts between 0.15 and 0.30% it minimises the susceptibility of the steel to temper embrittlement. In construction steels, it is used in the range 0.5–1.25% to give increased toughness and heat resistance. Higher contents (2–3%) have been used to aid corrosion resistance.

Copper (Cu). It is usually used at levels less than 0.5%. It improves resistance to corrosion in certain aqueous environments (e.g. sea water). In combination with chromium it has been used in 'semi-rust-resisting' and high yield strength steels.

Aluminium (Al). Aluminium is used to control the grain size of the steel during heat treatment and hot working. It is also used to deoxidise steel – such steels usually having good toughness because of the fine grain size.

Silicon (Si). Silicon is one of the principle deoxidisers used in steelmaking. Rimming steels usually contain less than 0.05% while, for fully killed steels, the range is 0.15–0.30%. The silicon level may be reduced if other deoxidants are used. Silicon slightly improves strength. Above about 0.50% it can lead to cracking problems in welding.

Titanium (Ti). Titanium combines readily with oxygen and nitrogen in steels. In this way it can improve the effectiveness of other elements, in particular boron.

Boron (B). This element, usually added in small amounts (0.0005–0.003%), significantly increases hardenability.

Niobium (Nb). Niobium lowers the ductile/brittle transition temperature and raises the strength of low-carbon-alloy steels. It improves grain size, retards tempering, increases elevated strength and because it forms very stable carbides it can decrease hardenability.

Vanadium (V). Addition of this element inhibits grain growth during heat treatment and also imparts toughness and strength.

Calcium (Ca). Calcium may be used to deoxidise steels and is used to help control the shape of non-metallic inclusions which in turn can improve toughness.

Other elements, generally considered as impurities, can also effect hardenability and weldability. However, there are times when they are added as deliberate alloying elements. The absorption of gases during welding can cause porosity.

Phosphorus (P). Phosphorus increases strength and hardenability but severely decreases ductility and toughness. It is generally considered detrimental.

Sulphur (S). Sulphur reduces hot strength and is very detrimental to transverse strength and impact resistance.

Nitrogen (N). This element increases strength and hardness but decreases ductility and toughness. In combination with aluminium it forms nitrides which can control grain size and thereby improve toughness.

Oxygen (O). Oxygen can drastically reduce toughness.

Hydrogen (H). Hydrogen can cause serious embrittlement leading to cracking.

The effects described above generally relate to the elements added in isolation. When added in combination the behaviour may be different, if only in degree. Fig. 6.4 shows the relative influence of some of these principal alloying elements on the weldability of steels from metallurgical considerations.

By consideration of these relationships, it is possible to conceive of a method of estimating hardenability of a steel of a specific

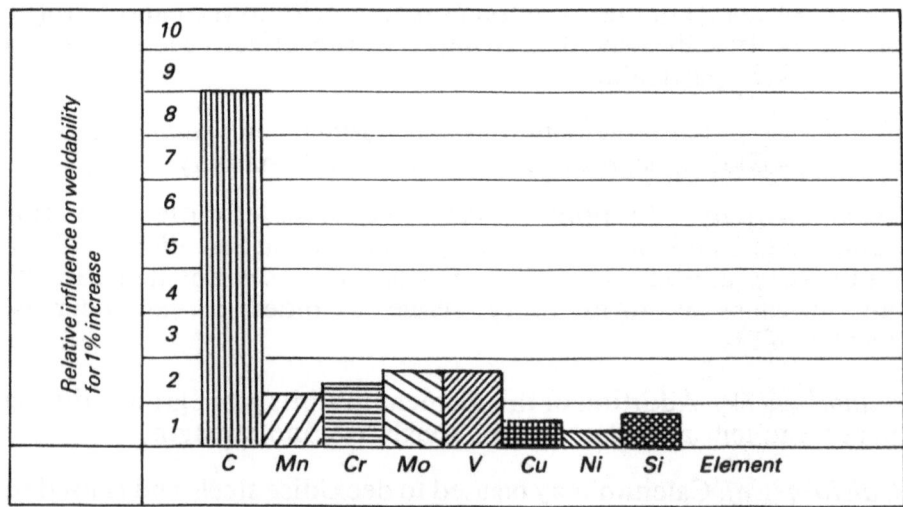

Fig. 6.4   *Relative influence of various additive and alloying elements on the metallurgical weldability of steels*

composition. The measure taken is called the 'carbon equivalent' (CE) because of the dominating effect of that element. The other elements are then weighted to relate their effects to the behaviour of additional amounts of carbon. One empirical formula for the carbon equivalent is as follows:

$$CE = C + \frac{Mn}{6} + \frac{Cr + Mo}{5} = \frac{Cu + Ni}{15}$$

where C, Mn, etc. denote the respective percentage composition.

Thus, given the composition of a steel, it easy to calculate the CE and to relate this to the results of welding trials and tests (e.g. measuring weld bead hardness). The batch of steel can then be accepted or rejected as suitable for welding according to the method and procedures used in the trials. Alternatively, the CE value will indicate the necessary changes to be made to the technology to ensure satisfactory welding (e.g. the use of preheat or the modification of welding parameters or sequence).

In a very much simplified and abbreviated form, Fig. 6.5 relates the CE of different families of steels to their weldability. This figure is intended only to give the reader an overview and does not contain sufficient information to provide a guide for actual applications. However, as a practical rule, for carbon or low-alloy steels of average thickness, it can be said that:

- Steels with a CE of less than 0.40 can be weldable without preheat.
- Steels whose CE is in the range 0.45–0.70 are weldable with preheats in the range 100–400°C; the required temperature increasing with CE and material thickness.

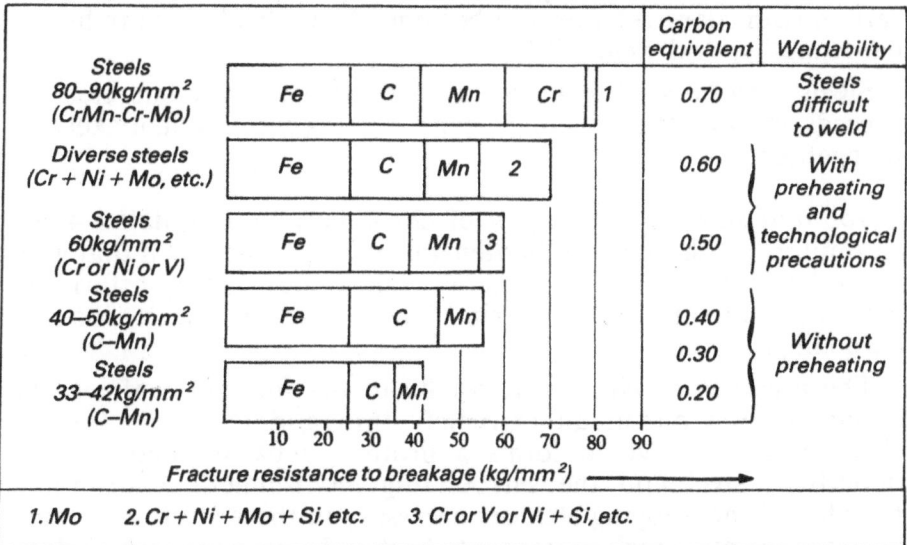

Fig. 6.5 Metallurgical weldability of different families of steels (CE values)

- Steels with a CE greater than 0.70 are difficult to weld, even with preheat.

It should be emphasised that the more difficult a steel is to weld the more important it is to follow the specified welding procedures and to obey the technological rules of welding.

The thickness of the metal being welded affects the rate at which heat can be conducted away from the joint and thus the cooling rate. This effect has been related to a carbon equivalence by SNCF (French National Railways) as a 'compensated carbon equivalent' (CEC) as follows:

$$CEC = CE + 0.00254e$$

where e is the thickness of the part being welded in millimetres. (Note that this relationship is only valid for a limited range of alloying elements.) Steels with a CEC greater than 0.45 are considered difficult to weld without preheat.

This equation can be used to estimate thickness which can be welded as follows: a mild steel, widely used for welding, which has a CE of 0.16 should not need preheating in thicknesses up to about 115mm ((0.45–0.16)/0.00254). The welding of such sections should also be considered from practical process considerations which will be discussed later in the chapter.

Characteristics of low and high alloyed steels

Steels are considered to belong to the class of 'low alloyed' when any one constituent element is greater than a threshold value. As alloy content increases, especially with the highly alloyed steels, the simple carbon equivalent relationship becomes less reliable for predicting hardenability and weldability.

Steels may be said to be highly alloyed when any one element exceeds 5%, with the principal elements being chromium, nickel and manganese.

Chromium steel. One category of chromium steel contains 4–6% chromium which increases hardenability to such an extent that with natural cooling from the austenite range they will transform to martensite rather than to ferrite. Such a steel is said to be air-hardening.

The presence of chromium results in the formation of, on the surface of the metal, an extremely thin and tenacious film of chromium oxide which forms a protective barrier between the underlying steel and the surrounding environment. Chromium, therefore, is an alloying element necessary in those steels which are required to have a high resistance to heat and corrosion. To be able to resist aggressive environments, the chromium should be at least 11%.

The rust-resisting martensitic steels contain from 11 to 16% chromium and 0.15–1.20% carbon and, as noted above, are air-hardening. These steels are designed for use in machine parts where creep properties are critical and high strength, corrosion resistance and ductility are required.

In the chromium range 17–27%, with carbon 0.08–0.35%, ferritic body-centred cubic structures are stabilised. These ferritic stainless steels are about 50% stronger than plain carbon steels and are used for decorative trim and equipment subjected to high pressure and temperatures.

Nickel steels. Nickel is used as an addition element to those steels requiring corrosion resistance and good properties at very low temperatures. When added in low concentrations it increases hardenability, while at high alloy contents it assists in the retention of austenite.

Industry uses three principal categories of nickel steels: those containing about 3.5%, those with 9% and those with 18%. The first group, of low alloy content (less than 5%), have a carbon content of about 0.15% and are used for service temperatures down to about −100°C. The second group has better toughness and ductility and is used at lower, cryogenic, service temperatures (e.g. the storage and transportation of liquefied gas).

The third group comprises a class of high-strength steels known as 'maraging' steels. The term derives from martensite age-hardening which relates to the hardening of a low-carbon martensite by the precipitation of intermetallic compounds. These steels are special in that this hardening reaction does not involve carbon. The age-hardening process can, for example, be explained by considering an 18% nickel maraging steel. The metastable diagram (Fig. 6.6) shows that when cooling this steel no phase transformation occurs until the martensite (M_s) temperature is reached around 250°C when only martensite is formed (even with slow cooling of heavy sections). Age-hardening is produced by heat treating for several hours at about 480°C, below the temperature at which the reversion from martensite to austenite commences (there is a thermal hysteresis in the austenite/martensite martensite/austenite transformations).

Maraging steels contain 18–25% nickel. Additions of other alloying elements (cobalt, molybdenum and titanium in particular) increases the yield strength to where it closely approaches the tensile strength (e.g. 2000 and 2050kN/mm²). These steels also have good low temperature toughness.

Austenitic manganese steels. Austenitic manganese steel, or Hadfield steel (after the inventor), contains 11–14% manganese and 1.0–1.4% carbon. This high manganese content prevents, when the steel is

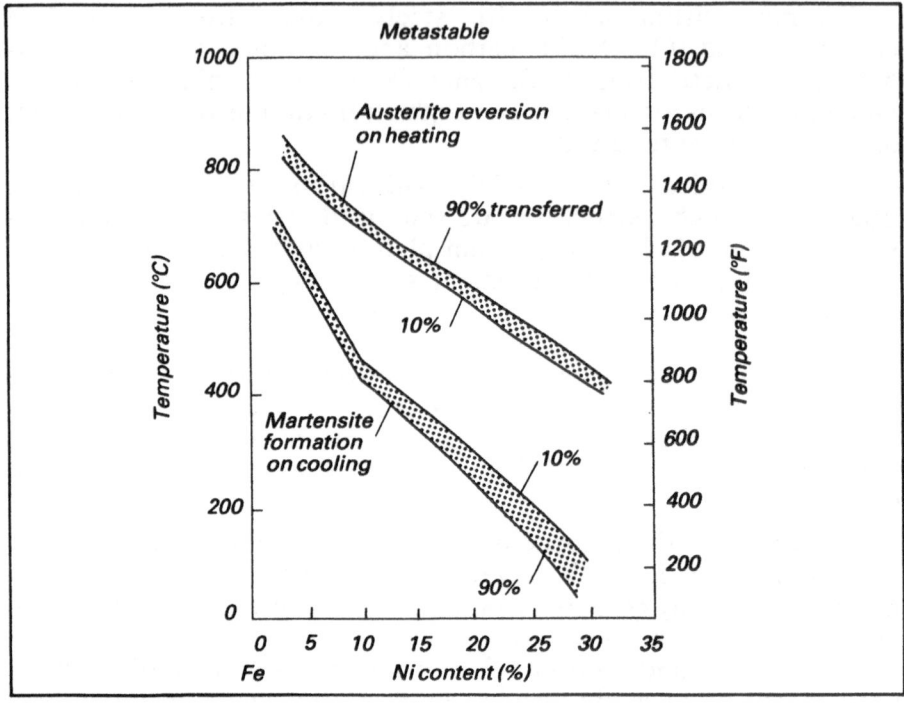

Fig. 6.6   *Metastable phase relationship in iron-nickel system*

quenched, the transformation of austenite into martensite. Deformation during service (especially impact) causes a hard martensite structure to form in the surface layer. As this martensite is a metastable structure, these steels should not be used as wear surfaces in, for example, rock crushers and railway frogs.

Austenitic chromium-nickel stainless steels. Chromium and nickel are often present together in steels resistant to corrosion and heat. Typically they contain 16–26% chromium, 6–22% nickel and up to 0.25% carbon. The nickel causes the retention of austenite at room temperature and therefore these steels are non-magnetic, thus providing a method of distinguishing them from other stainless steels.

A typical example contains 18% chromium and 8% nickel and less than 0.08% carbon. The more corrosive the environment the lower the carbon content should be. This is particularly important if the alloy is to be welded, as during welding there will be regions of the HAZ heated to between 425 and 800°C. At these temperatures, chromium carbides are precipitated in the grain boundaries, effectively depleting the chromium in this region. This results in a difference in material homogeneity with these grain-boundary regions being anodic with respect to the rest of the alloy. Under these circumstances there is a severe reduction in corrosion resistance.

This effect can be reduced by small additions of titanium, niobium or tantalum, elements which have a great affinity for carbon, thus 'stabilising' the chromium (hence the term stabilised stainless steel). Molybdenum additions of between 2 and 5% are also beneficial. The austenitic stainless steels have a high coefficient of thermal expansion and a low thermal conductivity. These factors lead to increased distortion, compared to carbon and carbon-manganese steels.

Elements other than chromium and nickel have similar effects and can be compared in terms of chromium and nickel equivalents (in much the same way as the carbon equivalent discussed earlier). For example:

$$Cr_{Eq}(\%) = 1 \times \%Cr + 1 \times \%Mo + 1.5 \times \%Si + 0.5 \times \%Nb$$
$$Ni_{Eq}(\%) = 1 \times \%Ni + 30 \times \%C + 0.5 \times \%Mn$$

Other expressions of these relationships include other elements such as nitrogen.

From the equations given, an austenitic stainless steel having a composition:

$$0.03\%C - 18\%Cr - 12\%Ni - 1\%Mn - 2.5\%Mo - 0.5\%Si$$

would have equivalents of:

$$Cr_{Eq} = 18 + 2.5 + 0.75 = 21.25\%$$
$$Ni_{Eq} = 12 + 0.9 + 0.5 = 13.4\%$$

Formulae of this type permit the calculation of alloy equivalents, which in turn enable the appropriate filler metal composition to be selected for the welding of the steel being studied. For example, to weld the steel indicated above one would choose a composition of approximately 21% Cr_{Eq} and 13% Ni_{Eq}.

The behaviour of stainless steels during welding can also be understood with reference to the Schaeffler diagram (Fig. 6.7) which relates the chromium and nickel equivalents to the metallurgical phases or constituents. With the aid of the diagram we can consider, for example, the welding of a carbon steel and the austenitic stainless steel listed above. In the diagram, the carbon steel composition is represented by point R and the stainless steel, and filler, by point S. If during welding, we assume that the dilution results in a composition equivalent to one-third filler and one-third stainless steel, then the weld metal composition would lie at point T on the line joining R to S (i.e. TS = RS/3, the point T being closer to S because the stainless metal is dominant). This point indicates that the structure will be predominantely martensitic which is an undesirable structure, for example, because of the increase in hardness. This structure could be

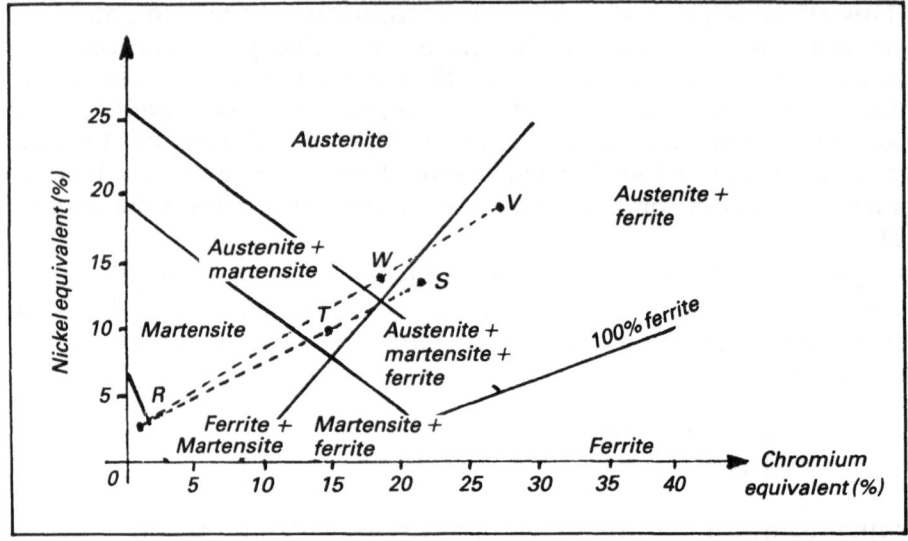

Fig. 6.7   Schaeffler diagram for welding of a rust-resisting austenitic steel and a
carbon steel

avoided by the use of a filler metal much richer in chromium and
nickel. If we consider a filler with chromium and nickel equivalents
of 27% and 18%, respectively (point V on the diagram), and assume
the same dilutions, then following welding the weld metal would
have a composition represented by point W which is fully austenitic.
The diagram indicates that:

- The filler metals for welding austenitic stainless steels should
  generally have chromium and nickel equivalents similar to those
  of the parent metal.
- When welding the austenitic stainless steels to carbon and low
  alloy steels the filler metals should have high chromium and nickel
  equivalents and dilution should be kept to a minimum.

Welding of carbon and low alloy steels
Weldability metallurgical considerations will be dominated by the
ability to weld without difficulties arising from:

- The formation of hard or brittle regions in the heat-affected zones.
- Grain growth and precipitation of carbides or similar embrittling
  phases.

These conditions are generally undesirable in that they can lead to
cracking and hence to fracture. These factors are related to the
intrinsic properties of the steel and are not caused as a result of the
process selected or the filler materials used. The effect is largely
governed by the mass of material, as this affects the rate of cooling
following welding. The rate of cooling will also be affected by the
heat input, both before and during welding. The thicker the material

to be welded the greater will be the mass available to absorb heat and the greater will be the cooling rate.

Thus to avoid these problems the relationship between joint mass, joint temperature and heat input (i.e. factors affecting cooling rate) must be considered.

Cooling rate can be slowed by raising the initial temperature of the workpiece above ambient (nominally +20°C), i.e. by applying preheat. Post-weld heating can also be used to control cooling rate (see also Chapter Seven).

The effects of preheat can be assessed practically following procedure welding trials by measuring HAZ hardness. Welding procedures or standards may specify a maximum HAZ hardness, with values in the region of 250–350 Vickers being typical.

We might define a weldable steel as one which, in thicknesses up to 35mm, does not require pre- or post-weld heating to avoid adverse metallurgical effects in the HAZ. Typically, carbon steels, even with sulphur and phosphorus contents of 0.05%, with less than 0.25% carbon would come within this category.

A moderately weldable steel is one which, in thicknesses up to 15mm, does not require pre- or post-weld heating. Steels weldable with this criteria are carbon steels with 0.25–0.40% carbon and low-alloy low-hardenability steels.

A steel difficult to weld is one which requires preheating for successful welding. Such difficult-to-weld steels include:
- Low alloy steels of the chromium-nickel and chromium-molybdenum type.
- Carbon steels with 0.4–0.6% carbon.
- High yield strength steels.
- Special steels, such as the air-hardening steels.

As noted earlier, the difficulty in welding these steels arises not just from the formation of hardened zones but also because of grain enlargement and carbide or similar phase precipitations.

Non-ferrous metals and alloys – Metallurgical considerations

Among the metals other than iron, the most commonly used are aluminium, copper, nickel, chromium and magnesium and their alloys. Both aluminium and magnesium and their alloys are considered as light metals.

Non-ferrous metals are generally more expensive than ferrous metals, particularly those based on chromium and nickel. Consequently, they tend to be used only when their special properties are needed, i.e. lower density, higher conductivity, higher thermal and electrical conductivity, etc. These properties are accompanied, for most applications, by sufficient mechanical strength.

Corrosion resistance and microstructure

One of the principal causes of corrosion in metals is the contact between dissimilar metals in the presence of a corrosive fluid. Different microstructures in the same alloy can also lead to corrosion in this way. Welding can cause such differences. Firstly, there can be differences in chemical composition which may arise due, on the one hand, to the loss of certain elements, say by vaporisation, and on the other hand, by the use of a filler metal (these are seldom chemically identical to the parent metal). Secondly, there are those changes, such as element migration to grain boundaries, which occur in the HAZ due to the welding thermal cycle. Since the use of non-ferrous metals is often associated with service in corrosive environments, there is a need to minimise these material/microstructural differences by appropriate treatments.

Influence of heating and cooling

The physical properties of a metal are important in their influence on weldability. To weld a metal with a high thermal conductivity naturally requires the application of a high/rapid input since the applied heat dissipates quickly. Another important physical property is thermal expansion. A high coefficient of thermal expansion translates into a significant shrinkage during cooling which can produce distortion and cracking (see Chapter Eight). The contraction on cooling also induces residual stresses which can lead to other problems such as stress-induced corrosion.

Welding thermal cycles can also destroy the properties developed by mechanical working or by heat treatment.

Hardening mechanisms

Consider first a pure metal which can be strengthened by reducing the path available for continuous crystal plane slippage by increasing the number of blocking grain boundaries, i.e. by reducing the grain size. It may also be increased by working, which may also reduce grain size but also increases the slip blocking areas (dislocations) within the grains. With alloys, two other classes of hardening are available: those based on solid-state reactions found in a relatively small number of alloy systems, and those based on alloy hardening which is found in all alloys.

Alloy hardening is a result of the alloying element going into solid solution and distorting the crystal structure of the parent metal. In this way, the slip of crystal planes is blocked. Alloy hardening also occurs if a second phase is formed, as the particles of the second phase also form blocks.

Solid-state (precipitation) hardening can be formed, for example, by eutectoid decomposition or by precipitation from solid solution. To produce the hardening effect, the alloy is first subjected to a 'solution treatment' to dissolve the maximum amount of the second phase in solid solution and then to retain this solution down to room temperature. This is done by:

- Heating the alloy to a high temperature, but below the temperature at which excessive grain growth occurs.
- Holding at temperature (sometimes for many hours) to allow solution to take place.
- Quenching in water to obtain a super-saturated solid solution.

The hardness after this treatment is relatively low, but is higher than for a slowly cooled alloy.

The full hardness of the alloy is developed by a 'precipitation treatment', during which the second phase is deposited as a fine dispersion throughout the structure. Natural ageing or age-hardening takes place at room temperature and it may take many days for the hardness to fully develop. The process can be speeded up by artificially ageing at temperature, albeit a relatively low one (e.g. 100°C).

Aluminium and its alloys

In historical terms, aluminium appeared relatively recently. It was first isolated at the beginning of the last century, and its properties, its silver-like appearance, its lightweight and the fabrication problems it posed, immediately made it a precious metal. Indeed, France's King Louis-Philippe, receiving a foreign sovereign at his table, honoured him with a table setting of aluminium, the other guests making do with gold and silver-gilt! The metal did not become an industrial material until the end of the nineteenth century when an electrolysis extraction method was developed.

The constructional value of aluminium can best be made through a comparison with steel. Consider, first, two plates of the same weight, length and breadth (Fig. 6.8). The thickness will be in the ratio of the densities, i.e. with the aluminium thicker by:

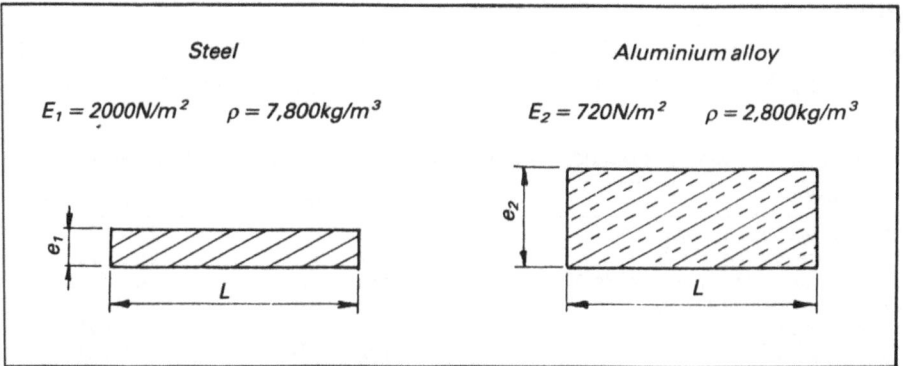

Fig. 6.8 Comparison between steel and aluminium alloy

$$E2/E1 = 7.8/2.8 = 2.8$$

The moment of inertia, proportional to the cube of the thickness, will be greater for the aluminium by 22 times.

The stress per unit area, when an equal force is applied, will be lower on the aluminium by 2.8 times. If, on the other hand, we consider the relative strengths if these beams carry equal loads, then for an aluminium alloy of 40kN/mm² tensile strength, the steel equivalent would be a 40 X 2.8 = 112kN/mm². Steels of this strength level are not commonly found in constructional grades suggesting a further benefit in using aluminium.

If the bending moment M is considered, the advantage may be even greater:

$$M = \frac{\sigma \, L \, e^2}{6}$$

where σ is the maximum surface stress in the beam, L is the width, and e the thickness of the beam. For a beam of equal width, and shape, as shown in Fig 6.8, and stress loading the aluminium plate can sustain a greater bending moment:

$$M_2/M_1 = \frac{\dfrac{\sigma \, L \, e_2^2}{6}}{\dfrac{\sigma \, L \, e_1^2}{6}} = \frac{e_2^2}{e_1^2} = \frac{(2.8 \, e_1)^2}{e_1^2} = 7.8$$

Table 6.1 presents a comparison between the major physical and mechanical properties of aluminium alloys and construction steels, with the ranges given taking account of effects of alloying and thermal treatments.

These factors indicate the value of aluminium alloys for various applications. For example:

• Its low density allows the fabrication of structures of high strength-to-weight ratios, as found in aerospace and naval

Table 6.1   *Physical properties of aluminium alloys and construction steels*

Properties	Aluminium alloys	Construction steels
Modulus of elasticity (kg/mm²)	7×10^3	21×10^3
Tensile strength (kg/mm²)	7–68	28–280
Thermal conductivity (cal/cm²/s/°C)	0.3–0.55	0.05–0.11
Electrical resistivity (Ω cm)	2.7–6	~60
Thermal expansion (cm/cm/°C)	28×10^6	15×10^6
Density (g/cm³)	2.63–2.81	7.8–7.9
Maximum fatigue resistance (R)	16	77

constructions in the body work of commercial vehicles and in railway rolling stock.

- Its high electrical and thermal conductivities has led to its wide use for electrical transmission systems and for heat exchangers as evidenced in domestic saucepans, etc.

- Its corrosion resistance has proved valuable, particularly in marine environments in the construction of transportion tanks and for pipework.

- Its strength and toughness at low temperatures has resulted in its use in cryogenic applications.

In using aluminium alloys, the designer must consider fully the materials properties and the way it is produced and fabricated, and not just transfer concepts directly from steel designs. For example, consider the modernisation of the Paris Metro where the carriages were made of steel. The new units were made of light alloys and, in the first instance, similar construction techniques to the original were used, and perhaps not surprisingly as the alloy was three times the cost of steel, were found to be more expensive.

The properties and methods of forming aluminium products were then taken into account which led to the carriages being manufactured from large extrusions, which resulted in economies in the use of labour, lighter units (by up to 50%) and reduced maintenance costs.

Welded aluminium alloy fabrications may be separated into two broad families:

- Fabrications using flat plate (up to 80mm thick) as the starting material (e.g. underground methane storage reservoirs, fuel gas containers, armoured vehicles, etc.).

- Fabrications making use of shaped thinner sections (e.g. rail carriages, military bridges, road transport vehicle chassis, etc.)

Aluminium alloys are relatively sensitive to temperature, showing a useful increase in tensile strength at temperatures below ambient but the reverse at higher temperatures, which generally limits the upper service temperature to about 200°C.

Aluminium is a chemically active metal which has a high affinity for oxygen. The oxide forms a tenacious surface coating (film) which contributes to the metal's ability to resist corrosive attack. These two factors can give rise to welding problems.

Pure aluminium and wrought aluminium alloys are designated by a four digit index system:

Alloy number	Major alloying element
1xxx	Commercially pure aluminium (99.00%+)
2xxx	Copper
3xxx	Manganese
4xxx	Silicon
5xxx	Magnesium
6xxx	Magnesium and silicon
7xxx	Zinc
8xxx	Other element

The other digits relate to alloy designations within the group.

These alloys can be grouped into two categories:

- *Non-heat-treatable alloys.* These alloys acquire their strength, from solution alloying elements, by strain hardening and by cold rolling. They fall into three characteristic groups: commercially pure aluminium, alloys with low manganese content and alloys containing magnesium. All are readily weldable and possess god corrosion resistance and toughness at room and cryogenic temperatures. Aluminium alloys are particularly useful because of their relatively high strength, even in the regions reheated by welding. The 1, 3 and 5 thousand series are considered weldable.

- *Heat-treatable alloys.* These alloys acquire their strength by mechanisms such as precipitation hardening which may be enhanced by cold working. These alloys contain, usually in combination, copper, magnesium, zinc and silicon. The commercial alloys most widely used comprise the groups containing: copper, magnesium-silicon, and zinc-magnesium. Generally the strength is obtained at the expense of weldability and corrosion resistance, and, with some alloys, reduced low-temperature toughness. The 2, 6 and 7 thousand series being considered weldable.

There are three widely used families of weldable alloys:

- *5000 series alloys.* They represent the most widely used family of weldable alloys for flat products. For example, alloy 5083, (Tables 6.2 and 6.3) is used for cryogenic applications.

Table 6.2   Chemical composition of alloy 5083

	Si	Fe	Cu	Mn	Mg	Cr	Zn	Ti	Al
Min.%	—	—	—	0.40	4.0	0.05	—	—	95.55
Max.%	0.40	0.40	0.10	1.00	4.9	0.25	0.25	0.15	92.55

Table 6.3   Mechanical characteristics of alloy 5083

R_m (MPa)	R_p (0.2) (MPa)	A (%)
270	120	16

- *6000 series alloys.* These alloys show good weldability despite a reduction of properties in the weld zone. Their essential quality lies in their formability, for example, by extrusion. The 6061 (Tables 6.4 and 6.5) and 6005 alloys are used in the manufacture of railway carriages.

- *7000 series alloys.* These alloys can be produced in both flat and shaped forms. They perform well in the welded state but their use is often specific as, for example, in military applications. Tables 6.6 and 6.7 give data on the 7020 alloy.

Filler metal
The choice of filler metal must be made bearing in mind the tendency of aluminium alloys to crack and for corrosion resistance to change following welding. The content of alloying elements is usually equal to or greater than the parent metal. The aluminium-magnesium alloys are widely used as filler metals because they show good strength as welded and they are not sensitive to cracking when used with a wide variety of base alloys.

Filler metals should be manufactured, handled and stored with great care to ensure the highest degree of cleanliness and absence of surface contaminents. Wires and rods are available in diameters ranging from 0.08 to 2.4mm in diameter.

Welding problems
Aluminium alloys are characterised by a high thermal conductivity (three times that of iron), a high thermal expansion (twice that of iron) and a low elastic limit – factors which can all contribute to

Table 6.4 Chemical composition of alloy 6061

	Si	Fe	Cu	Mn	Mg	Cr	Zn	Ti	Al
Min.%	0.40	—	0.15	—	0.8	0.04	—	—	98.61
Max.%	0.80	0.7	0.40	0.15	1.2	0.35	0.25	0.15	96.00

Table 6.5 Mechanical characteristics of alloy 6061

R_m (MPa)	R_p(0.2)(MPa)	A(%)
260	240	8

Table 6.6 Chemical composition of alloy 7020

	Si	Fe	Cu	Mn	Mg	Cr	Zn	Ti	Al
Min.%	—	—	—	0.05	0.9	—	3.7	0.08	95.27
Max.%	0.35	0.40	0.20	0.50	1.5	0.35	5.0	0.25	91.45

Table 6.7 Mechanical characteristics of alloy 7020

R_m (MPa)	R_p(0.2)(MPa)	A(%)
350	280	10

welding difficulties. Thus the use of a fast, powerful welding process is desirable to minimise the effects of thermal conductivity and to minimise distortion, and the use of large extruded sections can reduce the number of welds needed, in turn minimising the adverse effects of the low elastic limit (a low tolerance to restraint).

Other welding problems, for example, cracking, lack of fusion, porosity and metallurgical modifications, are influenced by the properties noted above and by other factors as discussed below.

Hot cracking. This phenomenon is similar to the hot cracks or tears which occur in castings. The weld metal cracks as the last liquid solidifies and as the weld is shrinking and contracting rapidly. The thin films of metal remaining become very weak and are unable to support the shrinkage stress and a crack develops. Those alloys with a long solidification range and low hot strength will be most prone to cracking.

Hot cracking may also occur in the HAZ where alloy migration to the grain boundaries has been sufficient to form low melting point constituents resulting in local melting and cracking during welding. Alternatively, the grain boundary alloys may simply lack sufficient hot strength to support the contraction stress, and again cracking occurs.

The restraint experienced by the weld has a significant effect on hot cracking, as does welding speed and welding mode. The following measures are useful in minimising hot cracking:

- Choose a filler metal, taking into account dilution, which will freeze rapidly (e.g. of eutectic composition).
- Use alloy additions which refine and strengthen the grain boundaries (e.g. titanium, zinc or beryllium).
- Increase welding speeds to reduce weld pool size, to increase freezing rate and thus to improve grain size.
- Use protective weld pool shielding gases, e.g. argon.
- Preheat to reduce thermal gradient to reduce shrinkage stress.
- Use a welding sequence to minimise stress build-up and residual stresses.

The use of robots for welding can aid in implementing some of these techniques.

Porosity. A survey of the literature shows clearly that the gas responsible for porosity in aluminium alloys is almost always hydrogen. The most common source of hydrogen is surface contamination of filler or joint preparation by hydroxides, hydrocarbons or oxides with absorbed water, all of which break down to release hydrogen. Water in the atmosphere can also be sucked into the weld region and give rise to porosity. The presence of

250ppm (parts per million) of hydrogen in the shield gas or 2ppm in the weld metal can lead to significant porosity. The pores range in size from the micro to 5mm or more diameter and are generally found aligned with the solidification isotherms.

Porosity can have an important influence on the properties of welded structures. Studies have shown that static strength and fatigue life falls in relation to the loss of cross-section caused by the pores. The effect on fatigue becomes more serious if the pores are near the surface or are surface breaking (possibly following post-weld machining). Unless they are gross, pores have little effect on Charpy impact values.

Lack of fusion. This occurs when the weld metal does not fuse with the adjacent parent metal or previously deposited weld metal. It can be caused by poor process and or parameter selection or weld pool placement within a joint. It can also be affected by the presence of the aluminium oxide which tends to persist as a film and which resists the wetting of solid metal by the molten weld pool. Magnesium and zinc can aggravate this situation. Oxygen can be introduced as a surface contamination or from the atmosphere.

Lack of fusion adversely affects joint strength, ductility and integrity, particularly when continuous and widespread.

Metallurgical discontinuity. Metallurgical heterogenities can arise from the thermal and solidification cycles induced during welding which are probably different from those of the parent metal. During welding, the base metal is brought at different points to temperatures, up to the melting point, one or more times. These cycles can affect both mechanical properties and corrosion resistance. The weld metal will consist of grains, often with second phases or alloy concentrations segregated in the grain boundaries. In the welded state, hardening (strengthening) comes mainly from solid solution effects. Properties equal to those of the worked or heat-treated parent metal can seldom be achieved. Furthermore, in the HAZ or in previously deposited weld metal, the thermal cycle can induce unscheduled ageing (or over ageing).

Physical considerations. During the preceding discussions emphasis has been placed on factors such as chemical composition, material thickness, and the rate of cooling as contributors to welding problems. Other factors that must be considered are the shape of the components and their stability, rigidity, etc.

It is easy to understand that for two parts made of sheets of the same thickness, but of different shapes, then it will be the one which shows the greater rigidity, by its dimensional stability, which will engender the greater shrinkage stress and therefore have the greater risk of cracking. With two parts of the same shape and external dimensions, but of different thickness, it will be that with the greater thickness which will be the focus of greater internal stress. Generally, all factors reducing distortion of a joint will tend to increase internal stress (see Chapter Eight).

Other considerations that may be mentioned, without it being possible to develop them further in the scope of this work, include the following:

- The basic shape and mass, as discussed previously.
- The distribution of components within the structure.
- The locking together of parts or the tacking of the structure which limits distortion: the last weld (closing weld) is always a delicate operation.
- The type of joint: a butt joint is generally more rigid than a fillet.
- The different factors defining the concept of the welded construction.
- The limit of elasticity of the base and of the deposited metal. The higher the value the more difficult it is to distort the metal. A steel with a high limit of elasticity is more prone to problems than an extra-mild one.
- The brittleness of the base metal: e.g. for equal shape and thickness, a cast steel part is generally more difficult to weld than one comprised of rolled sections (metallurgical problems also contribute to problems in this example).

Appreciation of the degree of distortion which will occur during welding is very important. However, the mechanism of the physically related factors affecting weldability are much more difficult to explore than the metallurgical factors. However, increasingly, these apparently abstract ideas are progressively being mastered by researchers.

To summarise, Fig. 6.9 presents, in highly schematic form, the weldability of a fabrication.

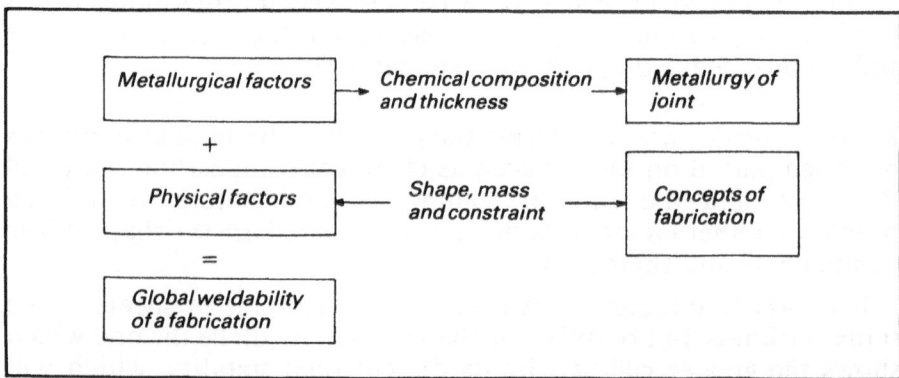

Fig. 6.9   *Highly schematic view of the overall weldability of a fabrication*

Chapter Seven

HEAT TREATMENT OF WELDED STRUCTURES

FOLLOWING the discussion of the previous chapter on weldability, it would seem appropriate to comment on the various thermal treatments available for improving the general quality of a joint. However, the objective of this text is to give a general outline, and since heat treatment is an engineering technique in its own right, it is recommended that the interested reader consult the abundant literature on this subject. The operation of these treatments – other than preheating, which, in fact, is not really a heat treatment (see below) – is relatively complex, involving the control of temperature and furnace atmosphere, and avoiding distortion and the oxidation of treated parts, etc.

Thermal operations before, during and after welding

Welding operations may be viewed as steps accompanied or completed by heat treatments whose objectives and operating conditions are very diverse. We can consider treatments such as annealing, tempering, normalising, hardening, etc., and those thermal conditioning operations for welding, such as drying, preheating and postheating.

Heat treatments may be carried out before or after the welding operation. They are intended to ensure either a relaxation of stress or a structural modification which favours the success of the welding operation, or an improvement in the mechanical characteristics necessary for good service behaviour. These treatments are generally applied globally to the parts before welding, to welded subassemblies or to the finished assembly. They may sometimes be applied locally when some level of residual stress is acceptable or when particular precautions concerning heating of the structure make global treatments impracticable or impossible.

Thermal conditioning operations are carried out immediately before, during, or after welding. They come under the category of what might be called 'welding precautions' and therefore constitute part of the welding procedure. Contrary to heat treatments, thermal conditioning operations are usually applied locally.

Tables 7.1. and 7.2. outline, respectively, the temperatures and objectives of thermal conditioning and heat treatments for metals and alloys currently used in industry.

Table 7.1   Temperatures of thermal conditioning treatments

Treatment	Low alloy steels (°C)	Cast iron (°C)	High alloy steels (°C)	Nickel and its alloys (°C)	Light alloys (°C)	Copper and its alloys (°C)	Object
Quenching			(Steel, 13% Mn)	(Hasteloy)			Avoid precipitations in HAZ
Warming/ drying	20–80						Avoid fracture by thermal shock
Preheating	100–350						Diminish effect of HAZ hardening
		500			200–300	300–500	Diminish level of residual stress
					200–300	300–500	Compensation of heat losses
Postheating	100–350						Prolonging transformation period and hydrogen diffusion

Table 7.2   Temperatures of heat treatments

Treatment	Low alloy steels (°C)	Cast iron (°C)	High alloy steels (°C)	Nickel and its alloys (°C)	Light alloys (°C)	Copper and its alloys (°C)	Object
Ageing			400 500	580 700	120 180		Structural hardening
Stress relief	550 750		850 900	815 1,180	300		
Tempering	550 750		725 (13% Cr)				Tempering of hardened zones
Intercritical	730 800						Partial recrystallisation
Annealing	850			815 1,180			Homogenisation, annealing or dissolving
Hardening	850 900						Increase in yield strength
Normalisation	850 900						Grain refinement
Re-annealing			1,000 1,100				Dissolving

Thermal conditioning

Cooling

Among thermal conditioning treatments, local cooling is sometimes desirable to limit the size of the heat field during welding. This is particularly the case with steels containing 13% manganese and with certain nickel alloys. With the latter, cooling restricts the formation of carbides in the heat-affected zone, which subsequently ensures a better corrosion behaviour. Cooling is generally obtained by spraying or sprinkling of water on the underside of the plates near the joints, usually during welding.

Warming/drying

Warming can be important when welding ferritic steels in very low ambient temperatures. For a long time welders had noticed that severe ruptures occurred when the welding was carried out on parts which were too cold. Experimentally they found that the risk was reduced when components were preheated only some tens of degrees centigrade (e.g. 30–50°C), hence the term warming. This low temperature has the effect of maintaining the steel above the ductile/brittle transition temperature, which for some construction steels is above 0°C.

Numerous codes and regulations prohibit welding when the ambient temperature is below a minimum (e.g. less than +5°C or 0°C). Warming is mostly required when welding steels of modest toughness. It is usually sufficient to warm the parts locally to between 20 and 50°C. (It may be necessary to give protection from air currents whose cooling effect may be intense.) This low temperature treatment is also sufficient to dry the joint surfaces and reduce significantly the susceptibility to hydrogen cracking.

Preheating

Preheating temperatures for steels range between 100 and 350°C, for cast irons between 500 and 600°C, for copper alloys up to 500°C, and for light alloys up to 300°C.

The application of preheating reduces the rate of post-weld cooling. For example, the application of preheat could increase the cooling period from 9 seconds to 17 seconds, all other operating conditions remaining the same. It is this effect which is used when welding hardenable steel. Local preheating may often be applied.

When welding cast iron, the preheating brings the part to a near plastic state which minimises expansion and shrinkage to avoid cracking. Preheating must be global (or the spread of heat must be at a constant gradient) to avoid the generation of dangerous stresses on either heating or cooling. (Note, these temperatures are maintained throughout welding and post-weld cooling is slow.)

Preheating of certain light alloys and cast copper may be necessary to avoid shrinkage cracking and must be carried out globally to as high a temperature as possible to minimise the thermal heterogeneities which produce shrinkage stresses.

In welding materials which are very good conductors of heat, the light alloys and in particular copper, preheating often must be applied to compensate for conducted heat loss and to facilitate the welding operation. In addition, on copper and its alloys, welding arcs are more stable when the parts are at high temperature.

The summary data given in Table 7.3 indicates that preheating facilitates welding operations on metals which are good conductors of heat, avoids thermal shocks and allows welding to take place above solid phase transition temperatures.

It helps, above all, when welding steel, to reduce the risk of hydrogen cracking by its combined effect on the phase transformations, the internal stresses, and on the evolution of hydrogen. There are some constraints, such as cost of preheating, discomfort to welders of hot components, and metallurgical embrittlement, which determine the preheating temperature.

Preheating has two objectives:
- To diminish hardening in the HAZ.
- To reduce the internal stress arising from constrained expansion and shrinkage.

Consequently it may be used for either reason, or for both. It does not modify the structure of the parent or weld metal, i.e. the

Table 7.3   Effects and consequences of preheating

Effects	Consequences positive	negative
Base metal temperature	Drying parts Work above transition temperature (brittle alloys)	Operator discomfort Risk of hardening if temperature is too high
Temperature gradient (reduction)	Improved operating weldability (metals and conductive alloys) Attenuated thermal shock	Enlargement of austenitic zone (steels) Enlargement of hardened zone (quench and tempered steels)
Maximum attainable temperature		Overheating Hardening (quench and tempered steels)
Cooling time	Between 800 and 500°C, attenuation or suppression of hardening Between M_s and M_f, less severe hardening Between θ_{max} and 100°C, diffusion of H^+, maximum stress differential	Hardening in the austenitic zone (quench and tempered steels)

metallurgical constituents (e.g. martensite, pearlite, austenite, etc.), their form (e.g. needles, lamellae, etc.), their size or the grain size, nor their distribution.

To summarise, preheating:

- Diminishes the brittleness of transformation zones.
- Reduces or eliminates the danger of crack formation.
- Diminishes distortion.
- Reduces internal stress.
- Facilitates the diffusion of hydrogen.
- And thus increases the ability of the joint to withstand service conditions.

As an operation, preheating is relatively simple to put into practice, and although sometimes onerous it justifies itself subsequently by ensuring that technical problems are avoided during or after welding.

In practice, it can be carried out by:

- Induction heating (typically for small parts, tubes, etc.).
- Electric resistance heating (widely used).
- Heating by oxyacetylene or oxypropane burners, etc. (e.g. for large parts that are difficult to handle).
- Heating in a furnace (for medium or large parts, recommended in the repair of cast iron parts).

As mentioned previously, preheating can be general or localised. When applied locally it must involve a zone equal to 10–15 times the material thickness of the part on both sides of the joint. The preheat temperature should be maintained until welding is completed. The temperature can be controlled:

- With the aid of heat-indicating pencils which are rubbed on the part. They are pencil-like rods, which have the property either of changing colour or of melting at a specified temperature. A range of temperatures can be measured by using different pencils to a precision of approximately 1% of the temperature being measured.
- By the use of thermocouples put in contact with workpiece. The temperature can be read directly from a galvanometer graduated in degrees centigrade. Thermocouples also allow the preheat temperature to be controlled automatically.

It can be difficult to determine the preheating temperatures required as the number of parameters to be considered is considerable, and includes the mass of the part. There are now computer programs which assist welding engineers in selecting the appropriate preheating temperature. Table 7.4 gives an indication of the preheat temperatures appropriate for a number of steel types.

Table 7.4   Preheat temperatures for steel types currently in use

Type	AFNOR designation	Base metal composition (%)	Recommended preheating temperature (°C)	General observations
Carbon steels	XC 18	C=0.15–0.20	Not necessary	
	XC 32	C=0.30–0.35	80–250	
	XC 38	C=0.35–0.40	100–300	
	XC 42	C=0.40–0.45	100–350	
	XC 48	C=0.45–0.50	150–350	
	XC 80	C=0.75–0.85	300–400	
Silicon steels	45 S 8	C=0.4; Si=1.6	To be avoided	Mediocre weldability
Chromium steels	100 C 6	C=1.00; Cr=1.4	600–800	Difficult to weld
	Z 12 C 13	C<0.15; Cr=12–14	500–600	Need for very slow cooling
Chromium-molybdenum steels	18 CD 4	C=0.15–0.20; Cr=0.8; Mo=0.20	200–300	Need for subsequent thermal treatment
	25 CD 4	C=0.25 ; Cr=0.8; Mo=0.20	250–375	
	42 CD 4	C=0.42 ; Cr=0.8; Mo=0.20	300–450	
Nickel-chromium steels	10 NC 6	C=0.12; Ni=1.4; Cr=0.90	200–375	Very difficult to weld
	10 NC 12	C=0.10; Ni=4 ; Cr=0.90	250–400	
	35 NC 6	C=0.35; Ni=1.5; Cr=0.90	350–450	Need for very slow cooling; sometimes thermal treatment
	30 NC 11	C=0.30; Ni=2.7; Cr=0.80	480–600	
	35 NC 15	C=0.35; Ni=4.5; Cr=1.70	500–650	
Nickel-chromium-molybdenum steels	30 NCD 8	C=0.30; Ni=2.2; Cr=0.55	350–500	as above
	30 NCD 12	C=0.30; Ni=3.0; Cr=0.90	400–550	
	36 NCD 16	C=0.35; Ni=4.0; Cr=1.80	500–560	
Manganese steels	Z 18 M 2	C=0.20, Mn=2.0	150–260	The heating of the part must be less than 250°C
	Z 120 M 12	C=1.25; Mn=12.0	Formally discouraged	
Corrosion resisting and resistant steels	Z 6 CN 18-10	C ≤ 0.05; Cr=18; Ni=10	No preheating	Steels of high carbon content may need preheating
	Z 6 CND 18-12	C ≤ 0.08; Cr=18; Ni=12		
	Z 6 CN 25-20	C ≤ 0.07; Cr=25; Ni=20		

Postheating

Postheating is particularly applied to steels sensitive to hardening, as it prolongs the period of structural transformation and hydrogen diffusion. However, it may produce a considerable tempering effect and this factor must be taken into account.

Post-weld heating has a two-fold effect in that it influences material behaviour in the weld zone and alters the overall distribution and intensity of the residual stress.

With certain steels there is also a need to control the interpass temperatures (particularly when multi-pass welding) to minimise the likelihood of hydrogen cracking. It may also be necessary to apply post-weld heating. The need for such treatments can be best assessed given a knowledge of the design stresses in the assembly and the composition of the materials.

Table 7.5 summarises, for steel, the effects of in-weld and post-weld heating in relation to the temperatures at which austenite will start to transform from martensite (M_s) and at which the transformation is essentially complete (M_f). The treatment may involve a unique temperature hold (the first three columns) or may involve a more complex sequence with several holds (fourth column).

Thermal treatments after welding

This section considers, for carbon and low alloy steels, those thermal treatments applied after complete cooling of the welded assembly. That is:

- Treatments at a lower temperature than A_1 (see the appendix to this chapter), seeking the relief of stresses and/or tempering.
- Intercritical treatments, between A_1 and A_3, seeking a softening and an improvement in ductility.
- Normalising or annealing treatments (heating to the austenite range) aimed at thermal regeneration.
- Hardening and tempering of the welded assembly (again heating to the austenite range).

Table 7.5   Post-weld heating: behaviour of steel as a function of temperature θ_p

	Temperature			
Effect	$< M_f$	$M_f < \theta_p < M_s$	$> M_s$	level $< M_{s'}$ level θ_R
Temperature homogenisation	●	●	●	●
Hydrogen diffusion	●	●	●	●
Stress growth suspension	●	●	●	●
Complete martensitic transformation	●			●
Interruption of martensitic transformation		●		
Non-martensitic transformation			●	
Tempering				●

Table 7.6 summarises these treatments and their effects.

Stress relief treatment
This treatment is basically intended to reduce or eliminate the internal stresses resulting from welding. It should be noted that it does not directly affect the hardness of the transformation zones, but minimises the danger, for example, of cracking or of the fabrication distorting during machining. Thermal stress relief reduces the strength level of the material (at the stress relief temperature), enabling the elastic strain energy to be released in small but significant amounts of plastic distortion. The temperature limits are 620–650°C, and may be said to be a high temperature operation.

The rate of increase of temperature must be progressive and should be slower the more complex the shape of the part: an average of 120–150°C/h is recommended. The duration of the treatment, or more exactly the time at temperature, should not be less than one hour; the basis of calculation is three to five minutes per millimetre of thickness with a maximum duration of five hours. The speed of cooling plays a considerable role; the more so the greater the mass of the construction (a maximum of between 100–150°C/h is recommended). When the temperature has fallen to 150–200°C, it is possible to relax the control of the operation, and the part, if need be, can be removed from the furnace.

Table 7.6   The value and effects of post-weld thermal treatment

Temperature	Effects	Consequences	
		positive	negative
$\theta < A_1$	Local	Tempering	Stress creation according to arrangement Local softening
	Global	Stress relief Tempering	Global softening Transition temperature of base metal Fissuration on re-heating
$A_1 < \theta < A_3$	After annealing	Transition temperature following annealing by complete austenitic transformation	Softening
	After welding and tempering	Transition temperature of weld metal and affected zone Stress relief	Softening
$\theta > A_3$	Local	Thermal regeneration of fused metal and HAZ	Risk of hardening Creation of stresses Local softening
Cooling air or furnace	Global	as above	Risk of distortion
$\theta > A_3$ Hardening and tempering		as above	Risk of heterogeneity of hardening between fused metal and base metal

In fact, to be really effective, stress relief must be carried out on the whole structure and therefore inside a furnace. The results of local stress relief, performed with, for example, the aid of oxyacetylene torches or burners, may be highly erratic and may even produce adverse effects. This is where the experience of the fabricator is necessary.

Stress relief is not always necessary and usually can only be economically justified in special cases. For example:

- When the fabrications are to form part of machines requiring dimensional stability, especially if the fabrication is to be machined.
- When the structure must be of high integrity and must withstand high stress levels (particularly fatigue loading), as encountered in pressure vessels and alternator supports.

The standards specify the conditions for applying stress relief treatments; BS2633:1987, for example, for Class 1 arc welding of ferritic steel pipework for carrying fluids states that:

- The maximum temperature of the furnace when the pipework is introduced shall not exceed 400°C.
- The maximum rate of heating shall not exceed, for steels other than the chromium-molybdenum-vanadium and chromium-molybdenum types, 220°C/h for thicknesses up to and including 25mm. For pipes of thickness t over 25mm, $5500/t$ °C/h or 55°C/h, whichever is the greater.
- The temperature and the time at temperature is related to the steel composition and thickness; e.g. for carbon and carbon-manganese steels (up to 0.25%C), 580–620°C for 2.5min/mm with a minimum of 30min.
- The rate of cooling, for steels other than chromium-molybdenum-vanadium and chromium-molybdenum types, shall not exceed 275°C/h for thicknesses up to and including 25mm and for pipes of thickness t over 25mm $6875/t$ °C/h or 55°C/h, whichever is greater.
- Below 400°C the parts may be cooled in still air.

The temperature that the fabrication is introduced to the furnace is limited to reduce thermal shock. This temperature must be lower the more complex is the fabrication or the greater is the range or number of different thicknesses.

The temperature and duration of treatment must be sufficient to ensure stress relief, while avoiding creep distortion of the fabrication under its own weight, and the degradation of mechanical characteristics, such as elasticity and toughness.

Finally, the speed of cooling must be limited in order to maintain through thickness homogeneity of temperature and to avoid the re-introduction of high residual stresses which can appear if this last, very important, condition is not respected. The need for such treatment is more evident the greater the mass and the more complex the shapes.

Stress relief treatment is generally only applied to stainless steels where there is a risk of stress corrosion (and stress corrosion cracking). The heat treatment cycles with these materials are more complex and aim to avoid prolonged holding at elevated temperatures which favour the formation of chromium carbides (leading to loss of solid solution chromium and loss of corrosion resistance).

Annealing

In this operation, steels are heated above the critical temperature to allow the grains to recrystallise, and then cooled slowly. During the soaking period, the time at temperature, the metal reverts to austenite. Slow furnace cooling then takes place, which results in the formation of 'soft' ferrite and pearlite structures.

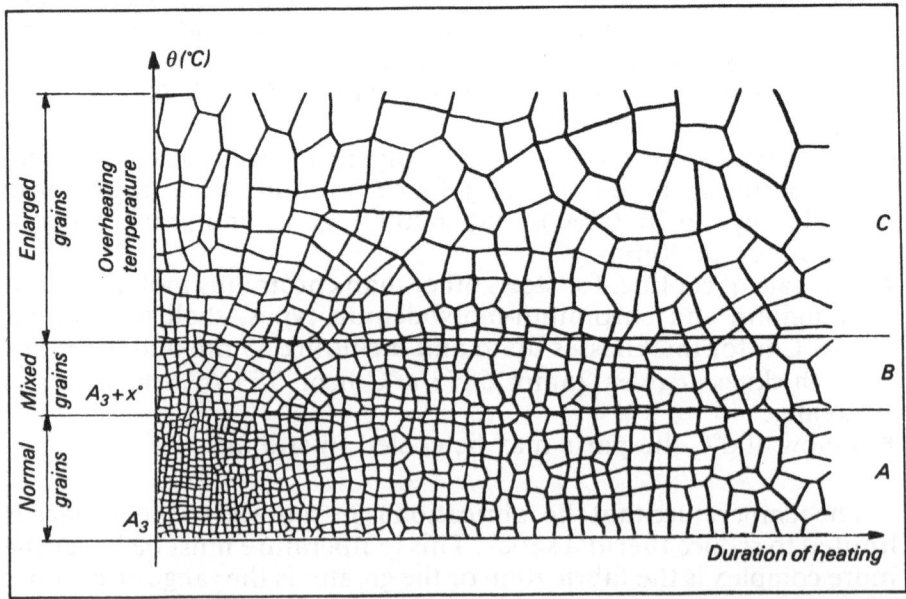

Fig. 7.1   Schematic view of enlargement of austenite grains as a function of temperature and duration of heating above the critical point. (Zone A, normal grains – an increase in grain size occurs when the heating time is very prolonged. The value of x is a function of the composition, the initial grain size, and prior mechanical working. Zone B, grains commencing growth – mixed structure, intermediate behaviour. Zone C, structure on overheating – the grain size increases rapidly the higher the temperature and the longer the heating time.)

Normalising

The steel is heated into the austenite range, until just fully austenitic, and is then air or control cooled. This cooling produces a finer, stronger and harder structure than the annealing treatment. The austenising temperature (A_3) should not be greatly exceeded, nor should the time at temperature, because the result will be excessive grain growth – the opposite effect to that aimed for (Fig. 7.1). Cooling rates should be slower with hardenable steels, i.e. those containing high carbon and or alloying elements.

To summarise, normalising:

- Principally grain refines – the fine grain structure improves the mechanical properties and improves metal working in operations such as forging.
- Suppresses stresses introduced by cold rolling and welding.

Normalising corrects many of the adverse features introduced by welding, e.g. hardened zones and residual stress are eliminated (or at least reduced) and toughness properties are improved.

Concluding remarks

The following points are worth noting:

- The need for heat treatment should be considered carefully, since to be effective it must be carried out fully and correctly.
- The beneficial effects of preheating (including component warming) are often underestimated.
- Annealing, normalising and stress relief treatments can improve service performance of welded joints improving the 'safety factor'.
- As normalising and annealing eliminate the weld-hardened zones, it might be thought that preheating could be avoided. This is not so as the preheat is, in part, used to avoid cracks forming during or soon after (hydrogen cracks) welding, and such cracks would not be removed by subsequent heat treatment.

Fig. 7.2 and Table 7.7 relate thermal treatments to the carbon content of steels.

Appendix: Points and intervals of transformation

This discussion follows FD.A02-010 of AFNOR (the French Standards Institute) concerning terminology. The definitions herein apply strictly only to carbon and low alloy steels. The nature and kinetics of the transformations and the temperature at which they occur can be profoundly modified by the presence of alloying elements.

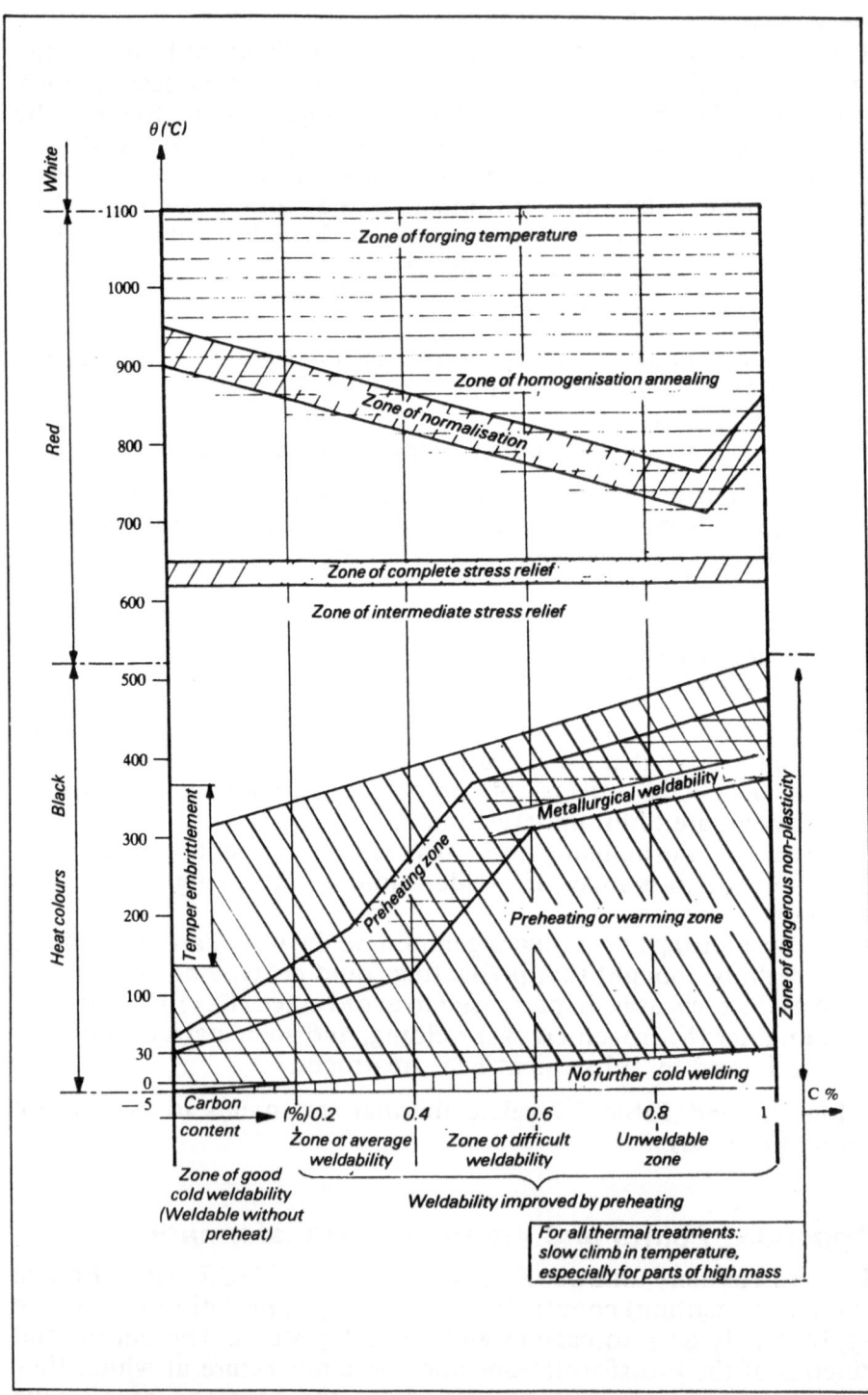

Fig. 7.2   Schematic of temperatures of thermal treatments of steels according to their carbon content

Table 7.7 Characteristics of the zones of thermal treatments shown in Fig. 7.2

Zone of homogenisation annealing – needs a subsequent treatment – little used in welding	The temperature is independent of carbon content Risk of grain enlargement
Zone of normalisation annealing – little used in welding	The temperature varies with carbon content
Zone of complete stress relief – used in welding	The temperature is independent of carbon content
Zone of preheating favourable to weldability – frequently used in welding	The temperature increases with carbon content (and mass is important)
Zone of preheating favourable to constructive weldability – frequently used in welding	The temperature increases with the mass of the part (and high carbon content)
Zone of no further cold welding	The low temperature is higher the more carbon content increases and the greater the mass
Zone of dangerous non-plasticity	(See Chapter Eight)
Blue-brittle zone	(See Chapter Eight) The steels are more brittle, therefore more crackable (apart from at low temperatures). It it necessary to avoid hammering or machining them

Transformation point or critical point of phase change

The temperatures at which there occurs, during heating or cooling of a steel, a change of phase are called transformation or critical temperatures.

The principal critical points are as follows:

- A_1 – temperature of austenite $\leftrightharpoons$ ferrite + cementite equilibrium. Austenite begins to form on heating and completes its decomposition on cooling, forming the ferrite + cementite eutectoid composition, i.e. pearlite. Related are: A_{e1}, the transformation equilibrium temperature (same as A_1); A_{c1}, the temperature at which the austenite begins to form in the conditions of heating employed; and A_{r1}, the temperature at which the transformation of austenite into ferrite and cementite (pearlite) ceases in the conditions of cooling employed.
- A_3 – temperature of austenite $\leftrightharpoons$ ferrite equilibrium in the case of hypoeutectoid steels, above which the austenite alone is stable and below which ferrite appears progressively. Related are: A_{e3}, the transformation equilibrium temperature (same as A_3); A_{c3}, the

temperature at which the ferrite to austenite transformation is complete in the conditions of heating employed; and A_{r3}, the temperature at which austenite begins to transform into ferrite on cooling.

- A_{cm} – temperature of austenite $\rightleftharpoons$ cementite equilibrium in the case of hypereutectoid steels, above which austenite alone is stable and below which cementite appears progressively. Related are: A_{ecm}, the transformation equilibrium temperature (same as A_{cm}); A_{ccm}, the temperature at which the dissolving of cementite is complete in the conditions of heating employed; and A_{rcm}, the temperature at which the precipitation of the cementite begins in the conditions of cooling employed.

- A_4 – temperature of austenite $\rightleftharpoons$ delta iron equilibrium.

- M_s – temperature at which the transformation of austenite into martensite commences on cooling.

- M_f – temperature at which the transformation of austenite into martensite is practically complete on cooling.

- The zones of transformation on cooling are sometimes designated by the following symbols: A_r, the area of pearlitic transformation; $A_{r'}$, the area of intermediate transformation; and $A_{r''}$, the area of martensitic transformation.

Critical interval (or critical zone)

The range of temperatures over which transformations occur lie between the A_1 and A_3 points, or between A_1 and A_{cm} points.

Transformations without change of phase

These may be described as follows:

- Magnetic transformation (Curie point) of ferrite – the temperature at which ferrite changes on heating from the magnetic state to the non-magnetic state, and vice-versa. This point of transformation is sometimes denoted by A_2.

- Magnetic transformation (Curie point) of cementite – the temperature at which cementite changes from the magnetic state to the non-magnetic state, and vice-versa. This point is sometimes denoted by A_0.

Chapter Eight

SHRINKAGE AND DISTORTION

IN THE shaping of materials, the greatest difficulty lies in obtaining and maintaining the exact dimensions and shape desired by the designer. For example, the founder – casting metals must take into account the global shrinkage of the metal, the restraint caused by the mould and the thermal heterogeneities occurring during cooling. Similarly, the carpenter must consider the moisture content of wood and the distortion which it causes, and the worker in concrete must allow for the shrinkage of concrete which can be restrained by the presence of metal reinforcing.

In fusion welding we face similar difficulties. Indeed, with many welded constructions the control of distortion presents the major problem to the fabricator. Experience, gained by observation, aided by certain proven rules enables distortion to be anticipated and its extent estimated. This knowledge forms the basis for selecting the correction procedures such as applying opposed preset, clamping (restraining), or back-to-back clamping (of similar components to balance opposed elastic distortion), or after welding (even if one practises corrective procedures) by rectification through heat shrinking.

It would seem, at first sight, that the task is easy, since most metals and alloys used are homogeneous, and better defined than the wood of the carpenter or the concrete of the builder. However, even the simplest welding operation involves a complex thermal cycle during which the metal is sequentially subjected to a high local temperature causing melting or softening (losing its rigidity), allowing it to give way in response to adjacent thermal expansion and, then on subsequent cooling, suffers the contraction related distortion which is difficult to analyse.

In total, welding shrinkage associates, in a complex fashion, physical distortion problems and metallurgical effects (e.g. the change of properties with phase changes or precipitation). In itself the analytical study of distortion problems constitute an arduous and thankless task. With the exception of the works of M. Gerbeaux of the Institut de Soudure (French Welding Institute), inspired by studies by Messers Walter in Norway, Okerblom in the USSR and Kihara in Japan, there are few bibliographical references on the subject (indicating a lack of attraction for Europeans to study this, nevertheless, important subject).

The distortions which occur during welding constitute a problem particularly subject to controversy. This stems, on the one hand, from the extremely complex nature of the phenomena and, on the other, from the fact that the corrective treatment often leads to an effect opposite to that desired.

At present, distortion is best avoided or controlled by the application of knowledge and experience acquired during previous work. It is therefore of great importance to observe the behaviour of all parts during welding.

There is no reason to consider distortion always as negative. The distortion stresses or contractions can also be used to bring structures into dimension by balancing one distortion against the other.

Expansion and shrinkage

Expansion
Generally, all heated metal bodies expand, with the result that dimensions increase in all directions. If you consider one point of the body as fixed (e.g. the centre), there is a displacement of the surrounding matter which can be defined by a system of three coordinates, whose origin is at the centre of the body.

For pure iron and mild steel, one can assume that the values of the displacements are the same along all three coordinates. It is therefore necessary to examine only one direction, and to this end, one of the dimensions of the sample is made much greater than the others. Expansion along this length is the measure of linear expansion.

Dilatometric methods are those used to measure the solid-state expansion of a metal (e.g. the method of Chevenard shows some important phenomena and allows the determination of the critical points of metals). Dilatometry relates dimensional changes to the temperatures at which changes of structural state occur during heating and cooling, (see Chapter Seven, Phase changes). These structural changes do not always occur at the same temperature during these two periods (Fig. 8.1).

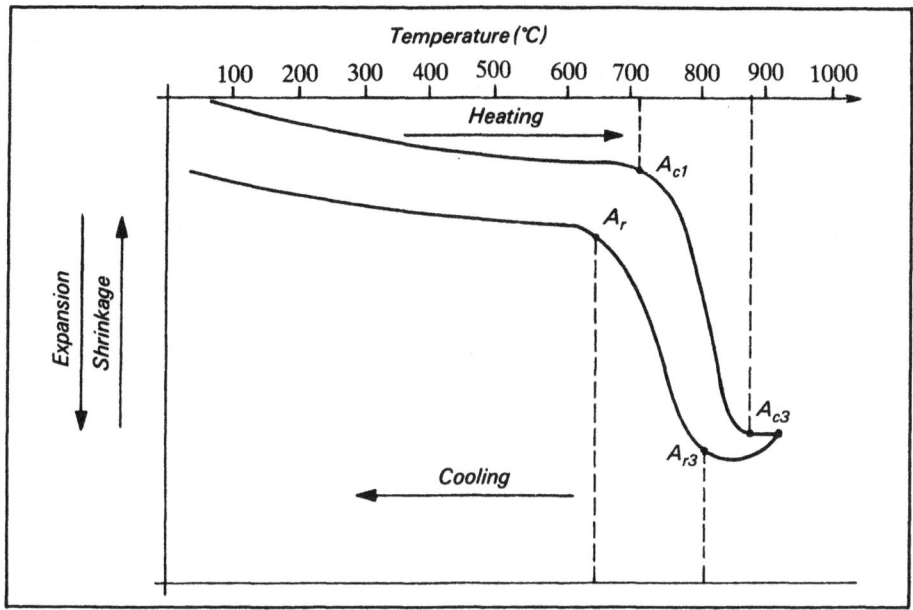

Fig. 8.1   Dilatometric effects when heating and cooling

It is possible to deduce, from the curves recorded by this method, the expansion coefficient at all temperatures. In examining, for example, the linear expansion curves derived from the diagram for mild steel (Fig. 8.2), one can deduce that expansion is almost linear up to about 720°C; above this temperature the ferritic structure begins to change into austenite, and since this has a higher density, the steel contracts. When the transformation is complete, near 870°C, the grains in the steel are entirely austenite and the expansion continues. The rate is greater because the expansion coefficient of austenite is higher than that of ferrite.

The reverse changes should occur again on equilibrium cooling. However, with faster cooling, the transformations can be delayed and the temperature of the critical point can be lowered, resulting in a thermal hysteresis effect. It is a function of different factors: speed of cooling, duration of heating, impurity and alloy content, etc.

To summarise, dilatometry allows the determination of:

- Thermal coefficients of linear expansion, i.e. increase in length per unit length per degree centigrade (cm/cm/°C).

- The temperatures at which crystalline modifications occur during heating and cooling.

Table 8.1 gives some coefficients of thermal expansion for different metals.

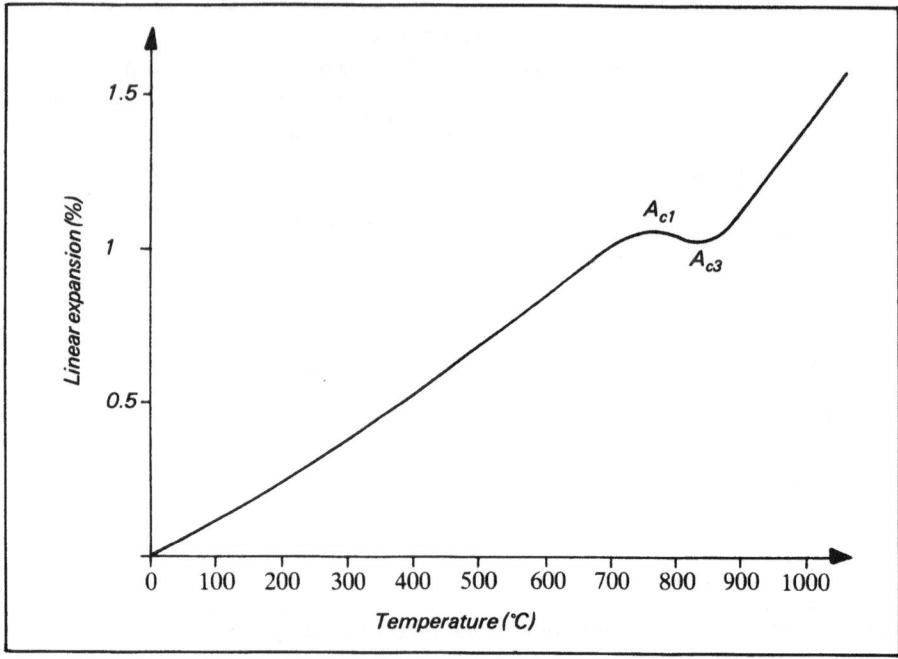

Fig. 8.2   Variation of the expansion of mild steel with temperature

Heating under constraint

Consider a bar of mild steel placed in a vice and heated uniformly
(Fig. 8.3). If the two clamps of the vice are fixed, the expansion of the
bar will be impossible, as is the case for a locally heated spot, such as
the weld bead, on the surface of a large plate.

If the bar (e.g. of 100mm length and 10mm width) is heated to a
temperature of 920°C and the distance between the clamps is kept
constant, the cross-sectional area will grow until the bar attains the
volume which the bar would attain if free and at the same
temperature. The cross-sectional area of the free bar, raised to 920°C,
will also be increased in size, but to a smaller degree because its
length will have increased by about 1.28mm. Thus, in restraining the
bar, the vice exerts a force equivalent to that needed to reduce the bar
length by 1.28mm.

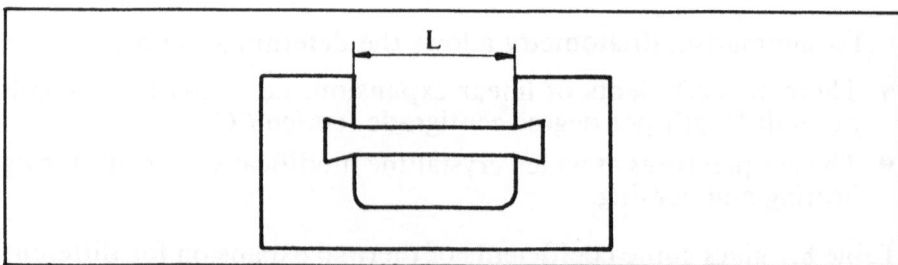

Fig. 8.3   Schematic of a metal bar gripped in a vice

Table 8.1 Coefficients of thermal expansion of various metals and alloys

Metals	Composition (%)	Thermal expansion coefficient			Elongation at the fusion point temperature (%)	State
		20–100°C	20–300°C	20–500°C		
Steel						
mild	C=0.17;Mn=0.4	11.8	13	14.2	2.2	Rolled and annealed
nickel alloyed	Ni=3.5	11.6	12.3			Annealed
with chromium	Cr=1.5;C=0.6	12.2	13.2	14.3		Annealed
	Cr=1.3;C=0.3	10	11			Annealed
austenitic stainless steels	Cr=18;Ni=8	17	17.5	18	2.8	
Aluminium	Al=99.7	23.7	25.6		1.6	Cast
Bronze	Cu=92;Sn=8	17.1	18.9			Cast
Aluminium bronze	Cu=89;Al=10;Fe=1	16.5	18.5			Cast
Bronze nickel	Cu=84;Sn=10;Ni=3.5	17.1	19			
Copper	Cu=99.9	16.6	17.6	18.6		Rolled
Grey iron	C=3.1;Si=2	11.1	12.2	13.2		Cast
Brass	Cu=70;Zn=30	17.8	19.1	20.5		Rolled and annealed
	Cu=60;Zn=39;Sn=1	19.8	21.2	22.5		Cold laminated
Monel	Ni=68;Cu=28;Fe=2	16.5	16.8	17.2		Cast
Nickel-chromium	Ni=76;Cr=19	12.4	13.6	14.7		Laminated
Nickel	Ni=99.4	13.1	14.3	15.2		Cast and annealed
Stellite	Co=80;Cr=20	14.1	15.2			

One can calculate the force the vice exerts on the bar as follows. For simplicity, assume that the bar is heated to 100°C above room temperature, The linear expansion of the bar is given by:

$$d = C \times \theta \times L$$

where C is the coefficient of linear expansion, θ is the elevation of temperature and L is the initial length. In this case the expansion coefficient is about 12×10^{-6} at the temperature attained, the lengthening of the bar $12 \times 10^{-6} \times 100 \times 100 = 0.12$mm. The formula for Young's modulus now allows us to determine the force necessary to lengthen (or shorten) the bar by 0.12mm.

$$F = (E \times dL \times A)/L$$

where F denotes the tensile force, A is the cross-sectional area of the bar, dL is the linear extension and L is the initial length. Young's modulus is constant for a given material; the value for the mild steel in this example is 21,000. On applying the formula one finds:

$$F = 21,000 \times 0.12/100 \times 10 \times 10 = 2,520\text{kg}$$

The equivalent stress level is 25.2kg/mm², which is close to the yield strength of the steel. We might then expect the metal to fracture if the temperature was raised much further. However, Young's modulus decreases with temperature, and furthermore when the limit of elasticity is reached, the steel is subject to permanent deformation which means that the stress remains very close to the yield value while the deformation continues (see Chapter Five). In fact, it is found, using this example, that if the temperature does not exceed 130°C, the yield strength is not exceeded and, on cooling, the bar will regain its initial dimensions.

Fig. 8.4 shows that the limit of elasticity of mild steel attains a maximum towards 300°C, decreases rapidly at around 500°C (to about 7kg/mm²), and falls almost to zero beyond about 800°C.

Shrinkage stresses
The stress field just described, developed when heating a restrained bar, is similar to that built up when a weld bead is deposited on a plate. The weld bead is not free to expand, being restrained by the plate. We have considered a restrained bar heated to 100°C, being subjected to internal stresses (of about 25kg/mm² by resisting an expansion of about 0.12mm) which on cooling, the limit of elasticity not having been exceeded, are relaxed and the bar returns to its

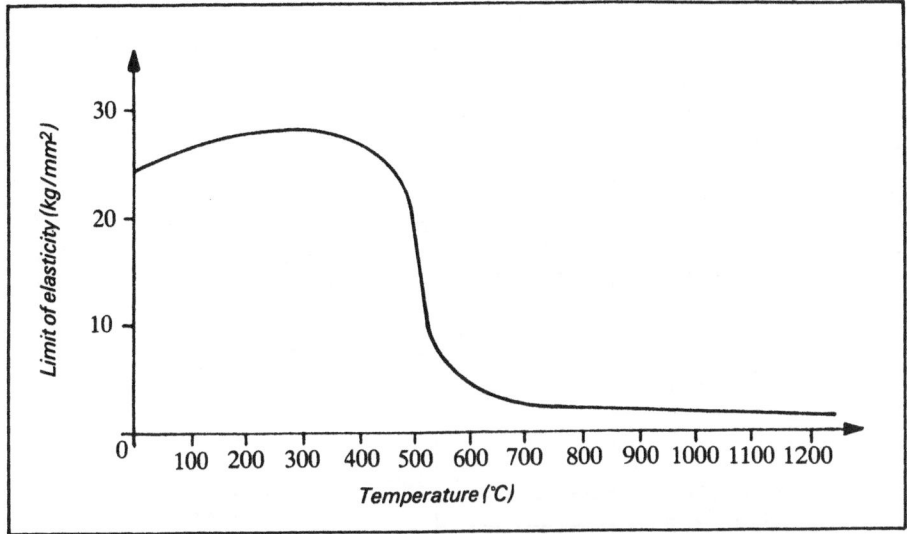

Fig. 8.4   Variations of the elastic limit of mild steel with temperature

initial state. We should now consider heating to, say, 800°C, and then a cooling to the ambient temperature. Figs. 8.5 and 8.6 detail the cycle. On heating the tensile stress increases to 300°C; thereafter it diminishes and by 650°C has become very small; at 800°C there is almost no stress, but the cross-sectional area of the restrained bar has been increased. On cooling, the now thicker bar being restrained, is unable to contract and a tensile stress grows, slowly at first and then rapidly as the temperature approaches 350°C attaining a maximum at around 250°C. At this point, the stress exceeds the elastic limit and the cross-sectional area of the bar decreases. The bar, after cooling, is therefore under tension. This behaviour exemplifies what occurs during welding, as welded deposits contain regions in tension due to such a contraction.

As well as the internal stresses resulting from the temperature changes and the permanent distortion, stresses can also arise from solid-state structural changes. Above about 540°C, such changes are readily accommodated because the limit of elasticity is low (Fig. 8.4), but if the transformation occurs at a lower temperature, due to rapid cooling, it takes place where the elasticity is higher and additional stresses develop.

Expansion and shrinkage play an important role in welding, but the study of these phenomena is complicated by the fact that the heating and cooling are localised. This results in the appearance of internal stresses which manifest themselves in shrinkage, folding, contracting, etc. To understand these effects during welding, we will consider three practical examples.

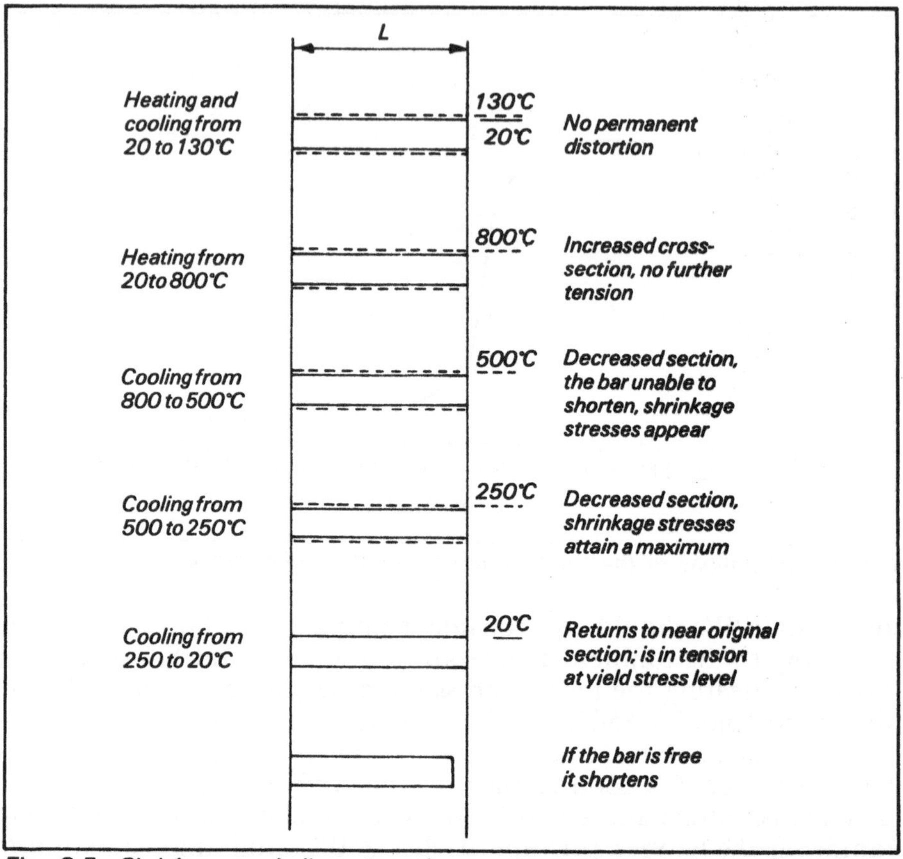

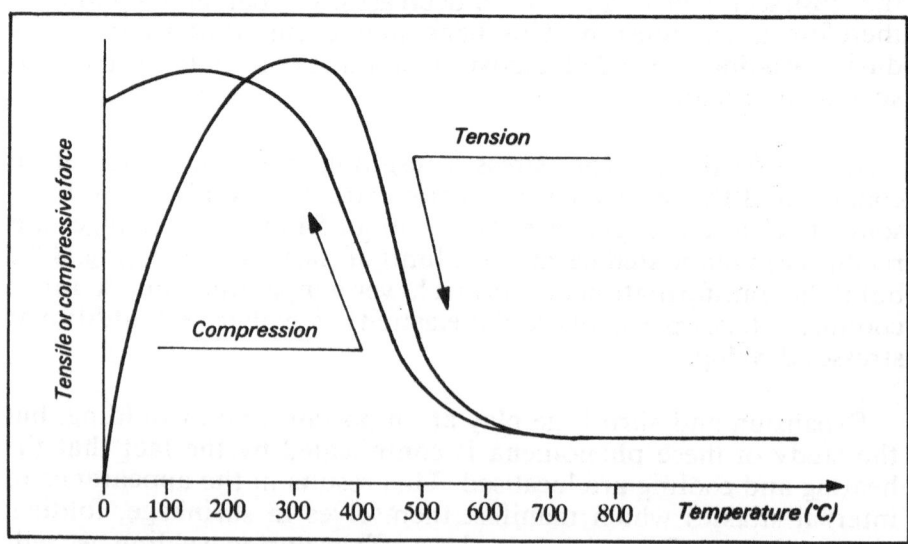

Fig. 8.5   Shrinkage and distortion phenomenon when heating and cooling a constrained bar (see Fig. 8.3)

Fig. 8.6   Variation of internal stress during the thermal cycle summarised in Fig. 8.5

Free expansion and shrinkage. All free parts that are homogeneous and heated uniformly expand, keeping their general shape, and on cooling regain their initial dimensions. The expansion and shrinkage are not constrained; the action is entirely reversible.

Inhibited expansion, free shrinkage. Consider a 'C'-shaped, 'constraint', part A, into which a block B fits snugly (Fig. 8.7a). On heating part B alone its expansion will be constrained in the vertical axis and it will become deformed (Fig. 8.7b). If the heating is carried on to a sufficiently high temperature (above 500–600°C for mild steel), B becomes subjected to significant compression forces *F* which cause it to plastically deform. On cooling, B, freely able to contract, keeps its new shape, with the result that its length becomes smaller than the initial length (Fig. 8.7c).

Inhibited expansion and shrinkage. If part B, as in Fig. 8.7, is an integral element of A (i.e. it will not be free to contract), the phenomena exhibited on local heating will remain the same (Fig. 8.8a). However, on cooling, B, endeavouring to contract, now experiences a tensile force supported by the branches of the 'C'. Once the solid has completely cooled, B is under elastic tension (Fig. 8.8b). This can be demonstrated by cutting through the middle of B, thus freeing (relaxing) the internal stress; the edges of the cut separate and the 'C' returns to its original shape (Fig. 8.8c).

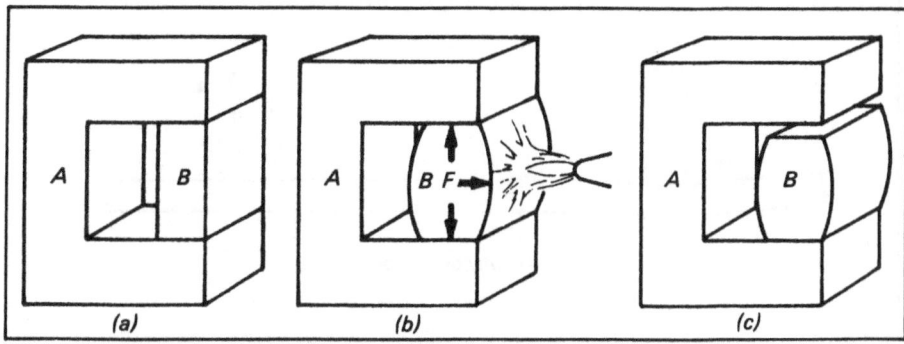

Fig. 8.7 Example of inhibited expansion and free shrinkage

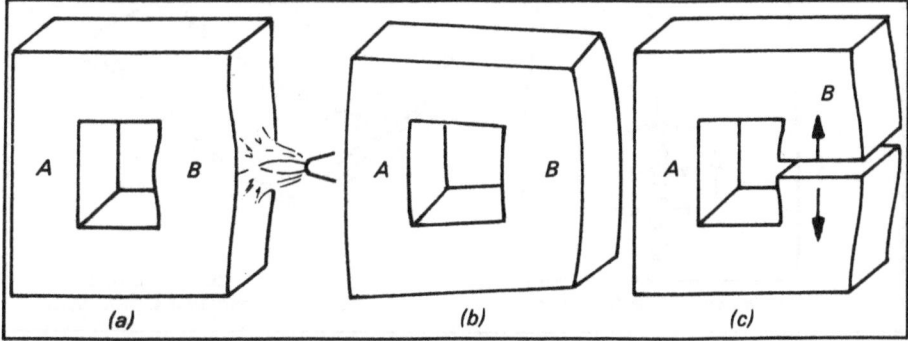

Fig. 8.8 Example of inhibited expansion and restrained shrinkage

Residual stresses and distortions

The above discussion shows that all localised heating with constrained expansion or contraction results in the appearance of (significant) internal stresses. In welded assemblies, it is generally considered that distortion due to shrinkage occurs in three directions:

- Perpendicular to the length of the weld (transverse distortion).
- In the direction of the weld (longitudinal distortion).
- Transverse to the surface of the deposit (angular distortion).

These distortions can occur simultaneously, but it is more convenient to examine them separately. Fig. 8.9 gives five examples.

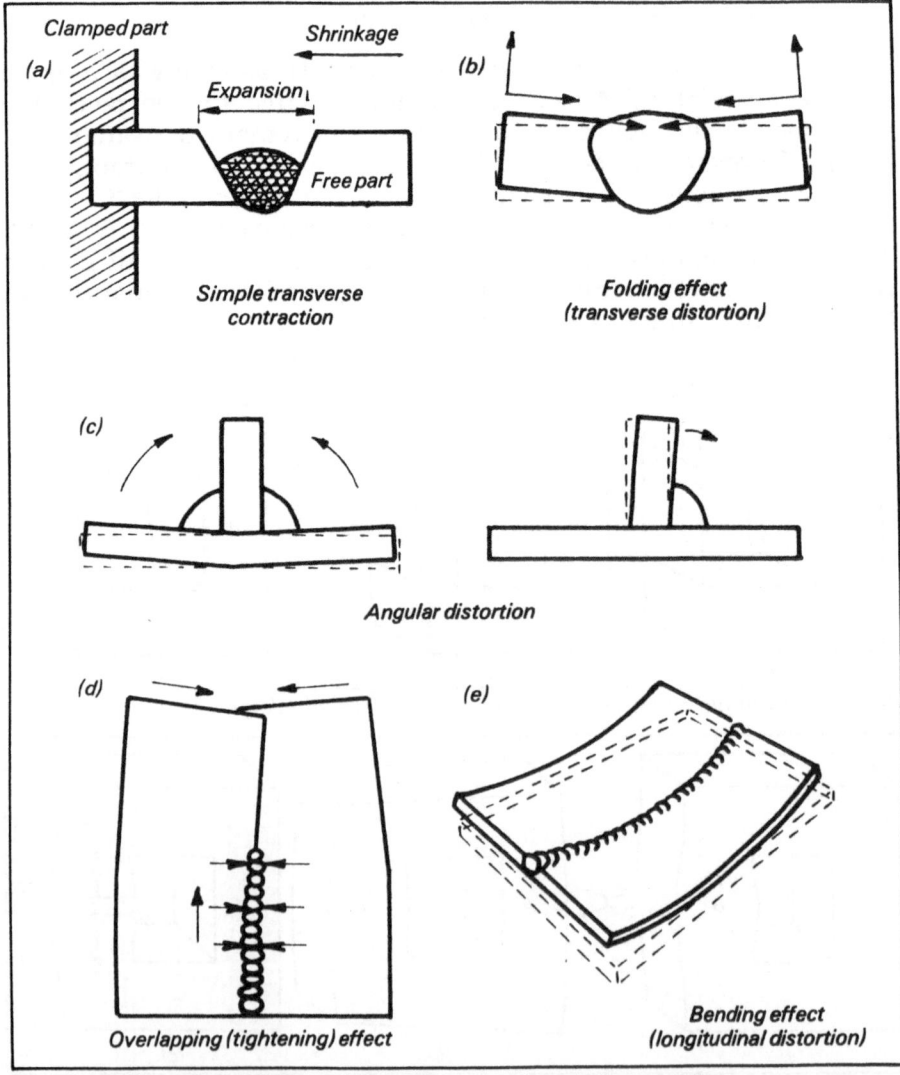

Fig. 8.9   *Examples of distortion types occurring during welding*

Transverse distortion

Shrinkage transverse to the weld depends principally on the cross-sectional area of the weld; the larger it is, the greater will be the shrinkage. For example, take the case of two mild steel plates of $200 \times 500 \times 20$mm thick, butt welded with a 5mm face gap. As welding progresses, the butt faces are melted and heat spreads into the plates, and naturally they expand. As the plates are now held or restrained by the plate already welded, the unwelded faces commence moving together. Now, in the area just behind the weld pool, the metal is still hot and plastic and presents little resistance to the thermal forces compressing this region. The metal shape changes, as for block B in Fig. 8.8, with the result that when the weld has finally cooled to room temperature the weld zone is trying to contract (or shrink) and is in tension.

The total shrinkage is due not so much to the contraction of the deposited metal but more to the contraction of the plates themselves. This shrinkage has been investigated by different authors, all of whom have noted a relation existing between shrinkage, the cross-sectional area of the joint and the thickness of the plate. Malésius gives the following formula for the transverse shrinkage of a joint, carried out in a single pass, on mild steel:

$$R_t = 0.1716s/e + 0.0121b$$

where R_t is the transverse shrinkage, s is the cross-sectional area of the weld, e is the plate thickness and b its root gap.

For the butt welding of steel plates more than 25mm thick using the manual metal arc process, Spraragen and Ettinger give different constants, for example:

$$R_t = 5.08s/e + 1.27b$$

For a 60° double sided V-joint with a root gap of 2mm, one could anticipate shrinkage to be: for plates of 15–30mm between 1.5 and 2mm; and for plates of 50mm thickness, about 2.5mm. When the plates are to be welded from one side, i.e. using a V-groove, then the shrinkage will be almost double that given above. For plates of thickness 8, 10, 12 and 15mm welded with a single-sided V of 60° inclined angle and a root gap of 2mm, the shrinkage is roughly 1.4, 1.5, 1.8 and 2.2mm, respectively. However, these figures vary widely and are influenced by other factors, for example:

- Welding process used.
- Residual stresses.
- Deposition rate.
- Diameter of the consumable electrode.
- Speed of welding (size of weld bead).
- Cooling period between passes.
- Sequence of deposition (welding plan).

Russel and Chihoski have studied the shrinkage of aluminium alloy welds in an attempt to explain the formation of certain cracks which occur at the centre of the bead. They considered first the distortions undergone by parts subjected to a welding thermal cycle, but without being joined. The expected distortion, calculated from the arrangement of isotherms produced by a heat source moving at a speed of 50.8cm/min and from the thermal characteristics of the metal, is as shown in Fig. 8.10a.

When the plates are being welded (i.e. being joined) under the same thermal cycle the solidifying bead constrains the thermal expansion occurring at the edges of the molten pool (point O, Fig. 8.10b). This constraint results in the build up of compressive stress at the points C_1 and C_2 and tensile stress at the points T_1 and T_2. These tensile stresses tend to separate the plates in front of the weld pool and cause a transverse contraction after the heat source has passed.

If the travel speed of the heat source is lowered, the isotherms become compressed and, all other things being equal, there is a greater temperature increase and the expansion is produced further in front of the weld pool. That is the stress C_1 becomes greater, and the corresponding increase in T_1 leads to greater separation of the plates. The stress C_2 moves with the heat source so that, after welding, only T_2 remains.

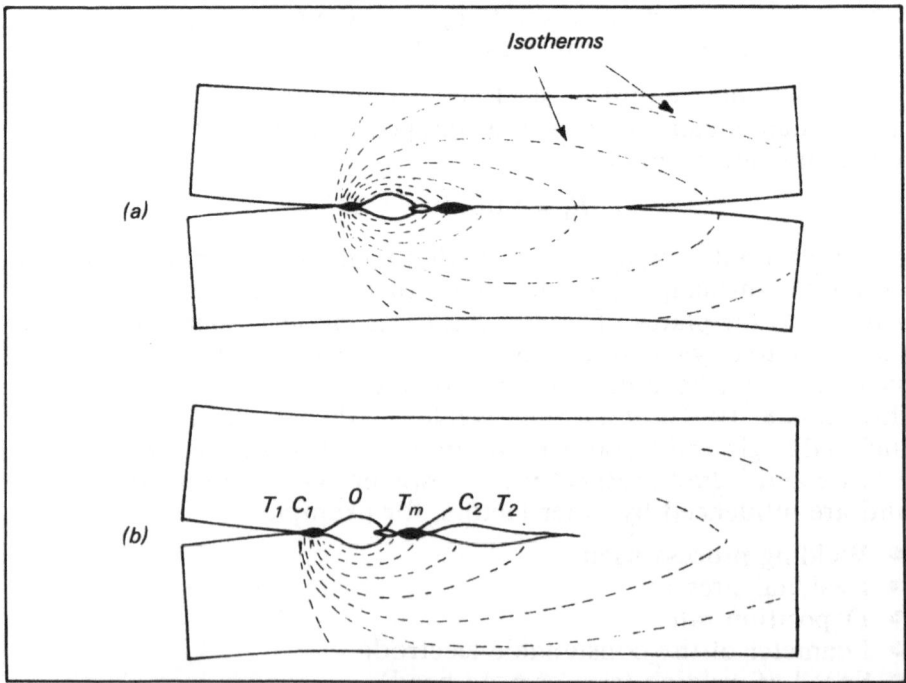

Fig. 8.10 *Shrinkage of an aluminium alloy plates: (a) unwelded plates experiencing a moving heat source, and (b) plates being welded with the same thermal pattern*

Fig. 8.11 shows the evolution of these stresses in the start region and in the weld mid-length as the plates cool. Fig. 8.11a shows the situation at a specific moment in time with Fig. 8.11b representing the situation a few moments later as the heat source approaches the plate ends. Here the stress C_1 dominates because there is less metal to conduct heat away and the metal becomes hotter. The result is that the plate ends have a tendency to close together. The residual stress field (Fig. 8.11c) when compared with those above are seen to be inverted (i.e. largely tensile) because of metal shrinkage.

As a whole, the parts of a welded joint have, along their length, been inhibited from expanding as the heat source passes, and thus the residual transverse stress arising from the shrinkage is tensile.

In summary, transverse shrinkage depends on the heat input over time, the site at which the heat is applied, the duration of the thermal cycle and the way in which the heat is allowed to dissipate.

Longitudinal distortion

As mentioned previously, longitudinal distortion is that which is exerted in the longitudinal direction (of the weld bead). This shrinkage tends to shorten the sheet at the site of the weld joint, with the effect diminishing as the thickness or dimensions of the resistant sections of the plate increase.

For example, a T-shaped steel girder assembled by fillet welds will, after cooling, curve into a 'banana' shape. This curving is produced

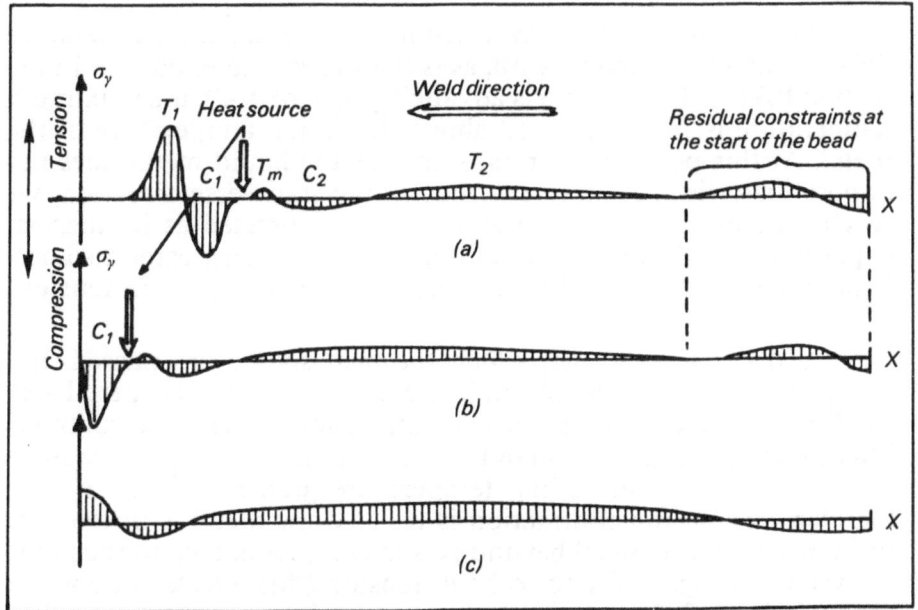

Fig. 8.11   Evolution of the distribution of residual stress in welding the assembly shown in Fig. 8.10b

by the longitudinal shrinkage of the two fillet beads, and is therefore a function of the cross-sectional areas of these beads. However, as the stresses will be carried by the total section of the T, the degree of shrinkage will diminish with an increase in this section. Values of shrinkage for such a T-section might range from 0.5 to 1.1mm/m.

Longitudinal shrinkage, then, is not negligible and can strongly distort certain assemblies. This is the case for box-section girders, where, to keep the girder straight, it is preferable to weld joints symmetrically. Weld shrinkage forces cause more distortion the further they are away from the neutral axis of the section.

To minimise longitudinal distortion, joints of long assemblies can be welded alternately and in an interrupted stepped sequence. Short lengths of weld are deposited moving towards the joint already made, such that each bead finishes at the beginning of the previous bead.

If the joint is asymmetrical, one would first preset the parts, the extent of the set being determined by experience. This form of distortion can be explained by considering the temperature and stress fields as illustrated by Fig. 8.12.

In Fig. 8.12a, the different parts of the plate are marked relative to the $x^l x$ axis, the axis of symmetry of the weld, and the $y^l y$ axis perpendicular to $x^l x$. Figs. 8.12b and 8.12c represent the distribution of temperature θ and the stress σ_x at various transverse sections.

In the section AA′ located in front of the weld, the temperature differential $\Delta\theta$ is practically nil, as is the stress differential, δx. In the section BB (the heat source is advancing toward AA′), the maximum temperature is obviously well above the fusion temperature of the metal. At this point, the stress is zero as the liquid metal does not oppose distortion. As we move out in the transverse direction, we reach a point where the metal is being compressed as its thermal expansion is inhibited by the cold metal located at a greater distance. From reaction, the metal located at distances $y > y_1$ is in tension.

At section CC′, through which the heat source has passed, the metal is in the process of cooling. Here, at $y = 0$, the metal has solidified, cooled, and is consequently endeavouring to contract. This endeavour is greater than that of the adjacent metal, because it experiences a greater falling temperature gradient. From $y = 0$, therefore, where the contraction is resisted the metal is in tension. From reaction, the metal beyond y_2 is in compression up to the point y_3, where it is again found to be in tension (this tensile zone is the continuation of that developed near the heat source which has moved outwards as the heat source moved away).

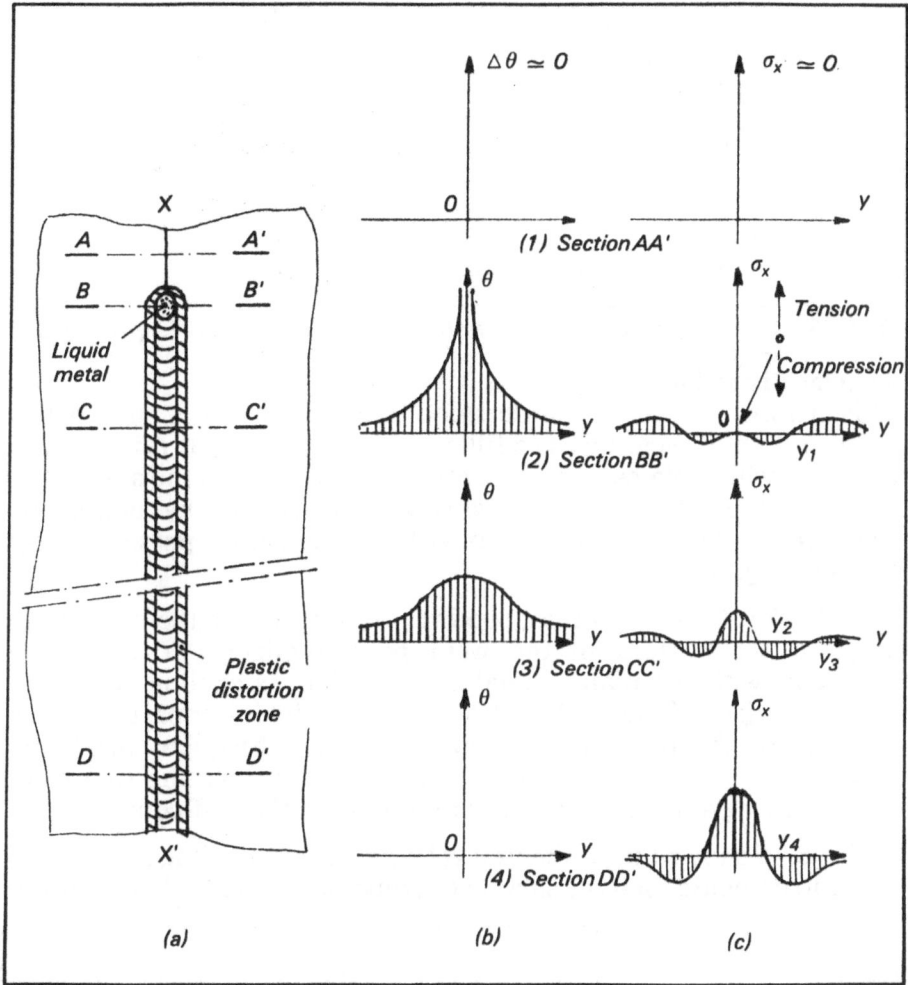

Fig. 8.12 *Analysis of the development of temperature gradients and stresses during a welding operation*

In the section DD′, located well behind the weld pool (in time and in distance), the metal has returned to near ambient temperature (i.e. zero temperature gradient). Here the residual stress is greatest at and around the origin, $y = 0$, where the greatest temperature elevation and hence contraction effect occurred. In this region, the longitudinal stress is tensile. At distances greater than y_4, the metal is from reaction is in compression. In all cases, the total of the longitudinal stresses along the transverse section is zero:

$$\int \delta x dy = 0$$

The distortions resulting from longitudinal residual stresses are typified by Fig. 8.13, which shows a narrowing of the plate.

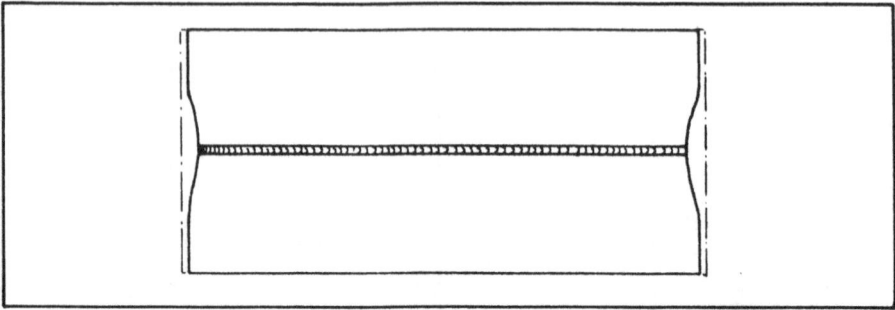

Fig. 8.13   Typical effect of longitudinal stress, resulting from shrinkage along the weld

Angular distortion

This distortion should be understood in its widest sense, and applies as much to butt welding as to fillet welding of plates. It refers to the tendency which plates have to contract (pivot) around the axis of the joint; and is due to a transverse shrinkage which is always acting off-centre in relation to the neutral axis of the weld pass or passes already formed (Fig. 8.14).

In the first bead deposited, cooling is more rapid in the base of the angle at the bottom of the weld pool, producing a transverse shrinkage with the plates remaining parallel. As cooling continues and the surface of the pool contracts, so the solid lower part serves as an axis for the pivoting movement. For a subsequent bead, the mechanism is similar and occurs more readily since the first bead already being solid moves the contraction further off-centre.

Fig. 8.15 shows that if the neutral axis and the cooling axis can be made to coincide then the shrinkage could be avoided. This situation

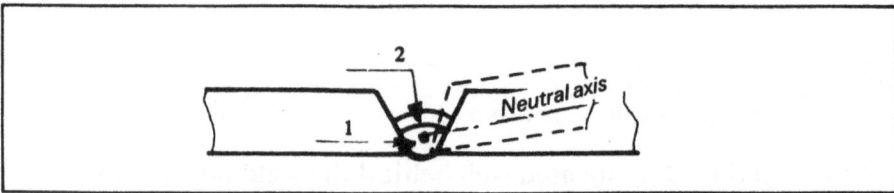

Fig. 8.14   Angular distortion due to transverse shrinkage

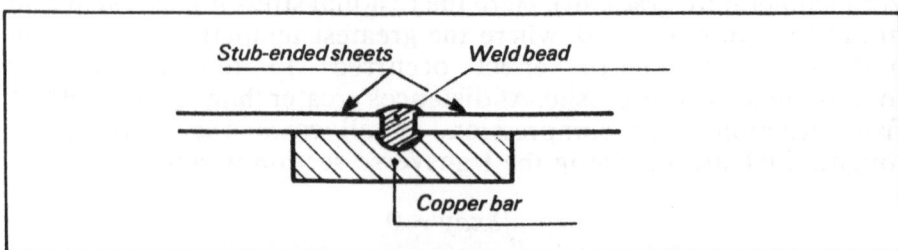

Fig. 8.15   Eliminating angular distortion by alignment of the neutral axis of the weld bead with the cooling axis (backing bar aids formation of parallel-sided weld)

presents itself when butt welding is carried out on a grooved copper backing bar (sometimes water-cooled), thus permitting very fast welding and the formation of a parallel-sided weld bead. (Electron beam, laser and plasma welds have low distortion because beads are narrow and parallel-sided.)

When welding sheet (thinner material), the joint can be achieved in one pass and there is practically no angular shrinkage. When butt welding plate, either prepared or unprepared, the more beads deposited the greater will be the angular shrinkage each time. Thus angular shrinkage is not a function of the cross-sectional area of the joint as in transverse shrinkage, but is proportional to the number of beads deposited. From the viewpoint of angular distortion, there is therefore an advantage to reduce the number of beads (Fig. 8.16). For the same total thickness of bead, the angular distortion is 6° 20′ for procedure (a) and 4° 20′ for procedure (b).

The reduction in the number of beads will also reduce distortion in butt, V- or J-shaped joint preparations. Even a joint of greater sectional area than another will give a smaller angular shrinkage if the number of beads is smaller.

It is important, however, not to forget that transverse shrinkage is also being simultaneously exerted, and that this is proportional to the cross-sectional area of the joint. Moreover, if it is the number of passes which has importance, the effect on the speed of deposition (i.e. the total welding time) should not be neglected.

It is seldom possible to determine angular shrinkage with sufficient precision to control distortion in thick assemblies. One could endeavour to make assessments by producing a reduced size assembly or even part of a joint. However, even if one takes the same quality plate, in the same thickness, one can seldom take account of the size and mass of the whole assembly, nor achieve the appropriate order and timing of the deposit. The total mass and overall welding time determine the cooling speed and distortion.

It is easy to understand that if one makes a second pass when the lower bead is completely cooled then angular shrinkage will be much

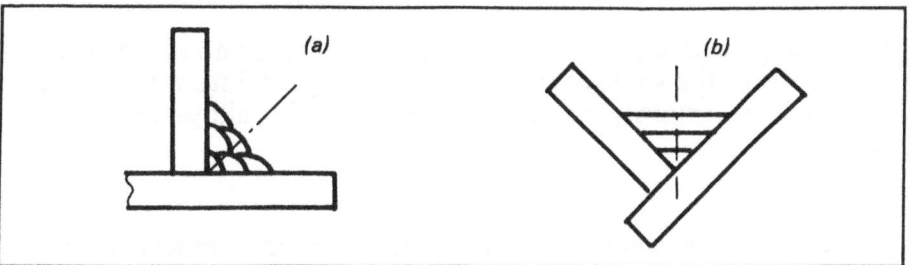

Fig. 8.16 Reduction of angular distortion by reducing the number of beads deposited (aided by changing welding attitude (b))

greater than if one welds on a hot bead which has not yet undergone its final shrinkage. Thus any tests for distortion should follow the procedures expected to be used on the full assembly. Such tests will tend to reflect the maximum angular distortion.

When the workpiece permits, final distortion can be minimised by giving the plates a set or counter-flex to an angle equal in value to the anticipated angular shrinkage.

Example of distortion of the base-plate of a steel girder
This example is given principally to provide some idea of the scale of shrinkage distortions.

During welding of a T-shaped or I-shaped girder, the fillet welds joining the central web to the base-plate (flange) each have their own angular shrinkage. In the case where the welds are not made simultaneously, the distortion effect is exerted.

From the tests and diverse results offered on this subject in the technical literature, the correlations of Braithwaite and of Ykman, a graph is presented allowing the prior calculation of the base-plate distortion (Fig. 8.17). This, however, is only given as a guide since, as noted previously, the many factors encountered in practice can modify the actual value of the distortion.

The theoretical calculation of the distortion is discussed below.

Longitudinal shrinkage. Few authors give quantitative values for this shrinkage. Blodgett proposed a formula for calculation of the longitudinal shrinkage caused when welding steel components possessing a certain moment of inertia:

$$R_t = 0.127sL^2d/I$$

where R_t is the total shrinkage (i.e. total vertical movement in mm), s is the total cross-sectional area, within the fusion line, of all weld beads (mm^2), L is the length of the fabrication, assuming welding of the full length (mm), d is the distance between the centre of gravity of the weld(s) and the neutral axis of the fabrication (mm), and I is the moment of inertia of the fabrication (mm^4).

Transverse shrinkage. Blodgett also proposed a formula allowing the calculation of the value of transverse shrinkage of butt, with as before the same uncertainty about the nature of the alloys to which it applies:

$$R_t = 2.54s/e$$

where R_t is the total shrinkage (mm), s is the transverse cross-sectional area of the weld bead (mm^2), and e is the thickness of the butt-welded parts (mm).

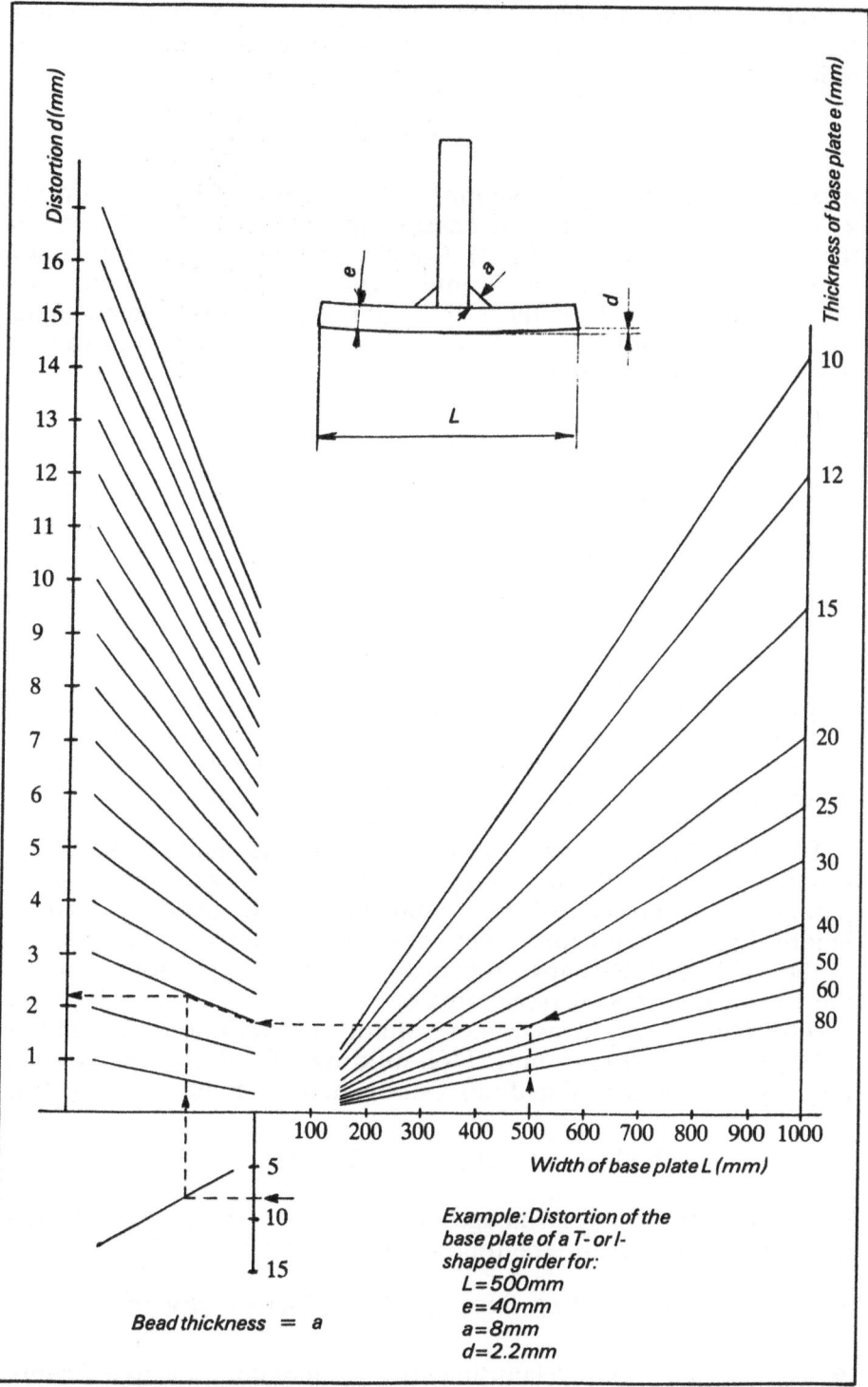

Fig. 8.17 Distortion of the base-plate (flange) of a steel girder following welding

Gilde established formulae for aluminium, rust-resisting steel and carbon steel. The formula giving the shrinkage of aluminium is as follows:

$$R_t = 2.04 \times 10^4 W/ev$$

where R_t is the total shrinkage (mm), W is the arc energy (arc current × voltage in watts), e is the thickness of the weld bead (mm), and v is the speed of welding (cm/min). Capel compares theoretical values calculated from the Gilde formula with the measured values of shrinkages resulting from the welding of two sheets of 50.8cm length, 12.26cm width and 6.35mm thickness, and welded in two passes by the TIG process. He found:

- First pass:
  - theoretical shrinkage: 2.5mm.
  - measured shrinkage: 2.2mm.
- Second pass
  - theoretical shrinkage: 1.6mm.
  - measured shrinkage: 1.1mm.
- Total shrinkage:
  - theoretical : 4.1mm.
  - measured : 3.3mm.

The measured shrinkage was lower than that calculated, but it must be noted that the joint preparation angles of 60° were low in relation to those generally used for aluminium (80/90°). The larger preparation angles would have led to higher shrinkage.

The Gilde formula seems from this the most adequate inasmuch as it takes account of the energy input and the welding speed – parameters which certainly play a part and which the other formulae ignore. The only ambiguity concerns the definition of e, the thickness of the weld bead. Perhaps a formula of the same type, but involving the transverse cross-sectional area of the fused part, would be more precise. Cline, using measurements made on aluminium copper alloy (2219-T8) after square butt TIG welding (direct current straight polarity with helium shielding), established the following formula:

$$R_t = 2.54 \ (e^{\frac{1}{2}} - 0.230) = 0.1 \ (25.4e^{\frac{1}{2}} - 5.842)$$

where R_t is the total shrinkage (mm) and e is the plate thickness (mm). Cline also made the following remarks:

- The speed of welding does not have a great influence on the transverse shrinkage.
- The use of a cooling/clamping arrangement does not significantly reduce the transverse shrinkage.

This formula and the comments are clearly limited to the particular alloy and process combination studied.

Angular distortions. Blodgett proposes the formula:

$$\Delta I = 0.51 \ Wa^{1.3}/e^2$$

to give the displacement ΔI (mm) under the effect of angular distortion of the extremity of the flange of welded steel I or T beams; W is the width of the flange, e its thickness, and a the leg length of the deposited fillet welds. (Note, this assumes the same fillet each side of the beam welds.)

Buckling defects during welding

In the production of welded assemblies, buckling defects often arise in front of the weld torch (Fig. 8.18). The defect is characterised by the gap i and the difference in misalignment (level) d.

Gott et al. studied this phenomenon. They made four observations:

- A pre-existing misalignment tends to increase during welding up to a limiting value.
- A pre-existing gap closes during welding.
- The phenomenon results in the need to vary heat input during welding to obtain a correct and constant penetration.
- Transverse shrinkage was dependent on the heat input and the degree of initial buckling.

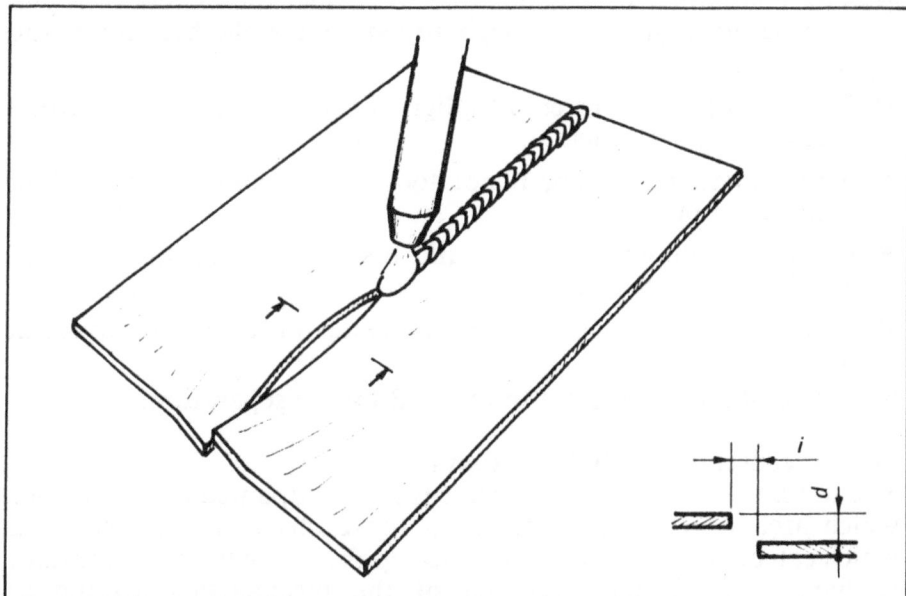

Fig. 8.18 Buckling defect during welding

Typically this defect can appear when producing the stub-end circular weld of a ring joint or depositing metal on a ring joint. As far as possible, the fabricator endeavours to achieve uniform fit up of the edges, but in practice there is always some misalignment. Experience shows that during welding the misalignment may more than double. Thermal asymmetries are thought to be a significant cause. These asymmetries can arise from imperfect metal-to-metal contact at the joint or with the backing support, or from a misalignment of the weld torch with the joint.

To investigate the effect of thermal asymmetries, Gott et al. introduced such an asymmetry by insulating one set of component clamps. They have shown that the part which is most cooled (clamped with no insulation) will have distorted the most when the weldment has returned to ambient temperature. They conclude that clamping (because of variable contact) does not always maintain the correct alignment of edges.

The findings of Gott et al. are summarised as follows:

- Generally, one joint edge increases in length more than the other.
- The edge which undergoes this increase is that which receives the most heat or which is the least well cooled.
- Thermal asymmetries resulting from poor clamping, arc misalignment, etc., are responsible for the differential expansion.
- In certain cases, notably circular welds, the increases of length translate into a difference in misalignment.
- As a consequence, it is possible to correct the effects of a pre-existing misalignment through adjustment to the heat input and cooling rate.
- The difference in level (misalignment) reduces the transverse strength of a weldment.
- A pre-existing gap can, if not too great , close in front of the welding torch.
- In closing, the edges can lap and disturb the control of the arc length and affect penetration.
- Variable gaps (and overlaps) will result in a bead of irregular geometry.
- The residual stresses are influenced by the separation.

Welded constructions in light alloys

The welding of aluminium structures is accompanied by distortions which are, in general, greater than those observed for steel, and which can prove costly to avoid or to repair. It is therefore important to have a good understanding of the mechanisms leading to distortion in order to master them.

Summary

To summarise, welding involves highly heterogeneous heating processes; it is essentially an anisothermal operation. The hot regions tend to expand, but are restrained by the resistance of the colder surrounding mass; they are thus compressed and plastically deformed. On subsequent cooling, the heated region endeavours to contract leading to the build up of shrinkage stresses.

It therefore follows that:

- When parts are 'free' to move ('unrestrained'), there is maximum expansion and the stresses are minimal.
- If the parts are 'clamped' or 'restrained' the expansion is minimal, and the stresses are high (often at yield point).
- The residual stresses arising during welding can lead to a diminution of the joint's cross-sectional area, reducing its load bearing capability weakens its mechanical properties. The stresses can also lead to cracking (crevice corrosion and/or stress corrosion), and possibly brittle fracture.
- The thermal stresses will be higher, and the risks of cracking higher, the lower the ambient temperature; high intensity welding is not permitted at low temperatures (e.g. below 0°C.)

More than for any other fabrications, residual stresses must be reduced in those destined for low-temperature service. These stresses may be reduced by relaxation treatments (such as reducing cooling rate).

Fig. 8.19 summarises the effect of joint volume on transverse shrinkage and internal stresses for welds in 10mm thick steel plate when in the 'restrained' and 'free' condition. Approximate numerical values are given.

Reduction of residual stresses and distortion

The observance of some elementary rules often suffice to reduce residual stresses and distortion to acceptable levels:

- Choose alloys having a capacity to resist or absorb distortion and resist cracking.
- Choose appropriate processes and parameters (e.g. narrow beads have low angular distortion).
- Balance welding to minimise distortion.

To reduce distortions it is possible to act before, during or after the welding operation. Among these methods, the most important are those used before welding.

Precautionary measures

These include the use of correct clamping or jigging, setting of joint angles, preheating and, above all, the establishment of an optimal welding sequence and welding speed.

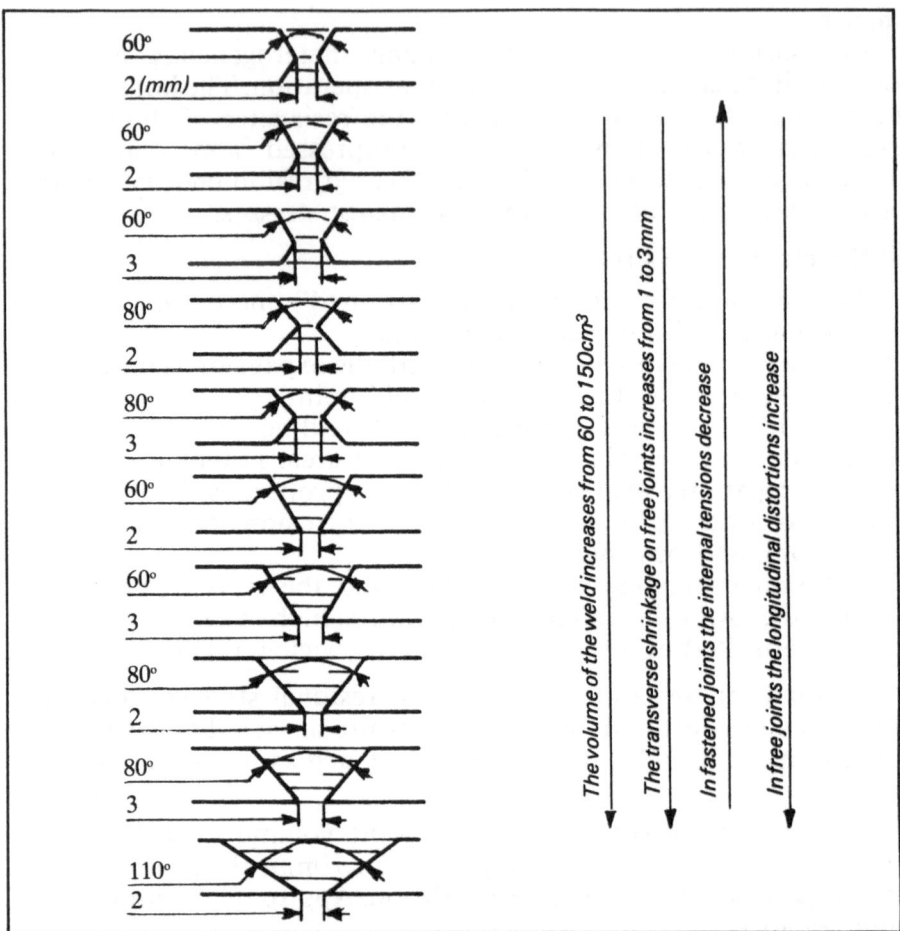

Fig. 8.19   Schema of the influence of fused metal volume on shrinkage and residual stress and distortions in both restrained and unrestrained joints

Jigging. In general, the welding of parts with a thickness of less than 5mm requires jigs to position and support. The jig system often incorporates a backing bar which supports the weld pool (often made of copper and water cooled). The jig must permit the precise and repetitive placing of the parts to be welded. It must also allow easy 'unloading' after welding (often difficult as a result of component contraction or distortion). The jig must also allow good access to the joint and, particularly if associated with a robot, should have one or more rotations to permit as many joints as possible to be welded in the flat position. Among other things this allows faster welding speed and results in lower distortion.

Presetting. This preset, made prior to welding, is of equal value but in the opposing direction to the anticipated distortion. It is a useful technique (where it can be applied) and is useful when automatic welding, where the welding parameters are reproducible. Fig. 8.17 can be used to estimate the distortion and thus to indicate the preset required.

Preheating. This method is mainly designed to prevent the formation of a hard brittle zone. However the reduction in cooling rates also reduces internal stress (see also Chapter Seven).

Welding sequence. For all welded assemblies one designs a 'welding sequence'. The plan indicates the order and method of executing the different welds with the object of minimising distortion and of obtaining an assembly low in residual stress. Whatever the process, the material(s), the size or the complexity of the assembly, there are three basic rules:

- Choose a welding sequence such that transverse shrinkage of each weld is minimised. Where the weld is highly restrained, then if possible choose, for example, an overlap joint in preference to a fillet, or a stepped welding sequence, or endeavour to design a more flexible assembly.
- Break the structure into subassemblies. Commencing with the detailed assembly, combine them progressively into larger subassemblies and eventually into the final product. For each assembly prepare suitable jigs and at each stage compensate for distortion.
- Use symmetrical or approximately symmetrical arrangements to form symmetrical subassemblies. Furthermore, symmetrically weld assemblies in order to maintain equilibrium and straightness.

These rules should be varied in relation to the demands of the weld design, the quality required and the actual working options.

Many welded structures can be self-jigging if the welding sequence is well planned. In other cases, tack welds are necessary to hold the parts correctly until the final welding is achieved. There exists a basic principle which is to carry out welds in reverse order of length; the shortest being performed first, allowing a good distribution of stresses and preventing the rupture of tacks. There is also value in spreading the welds in different parts of the whole assembly (as geometry permits).

The welding sequence should ensure:

- *Distribution of heat from welding.* This can be achieved in a number of ways:
  - by welding symmetrically around an axis or a spot.
  - by building up welds in short lengths (i.e. a blocking sequence).

– by back stepping (make weld in short lengths, the end of the second being at the start of the first, etc.).
– by opposed blocks (i.e. welding on opposite sides of the joint).
– by the use of copper heat sinks.
– by the type of pass (e.g. stringer or weave, Fig. 8.20).
Note that the multiplicity of passes improves the mechanical characteristics (construction security), and that the multiplicity of narrow passes increases angular and/or transverse distortion; that of wide (weave) passes increase longitudinal distortion.

- *Optimise speed of welding.* There exists an optimum speed for which distortions due to transverse shrinkage are minimal. If one butt welds two plates without tacking, they pinch together as indicated in Fig. 8.21a when the welding speed is very low (e.g. 20cm/min). If the speed is very high (e.g. 3m/min), the sheets have a tendency to open (Fig. 8.21b). At a certain intermediate speed, to be determined experimentally, the edges remain parallel. Welding with such an optimum speed is only practicable with an automatic machine or with a robot.

- *Control clamping.* As indicated above, the preferred procedure is to weld elements of an assembly in an order that allows most welds to be made unclamped. This contributes to a considerable reduction of stress and is highly recommended for welding parts made up from brittle materials (e.g. castings).

Note, these precautionary measures can only be perfectly applied by robot welding, which from this point of view becomes the most rational means of fusion welding.

Methods which can be used during welding
Masubuchi describes a study carried out by the Harvey Engineering Laboratories for NASA's Marshall Space Flight Center concerned

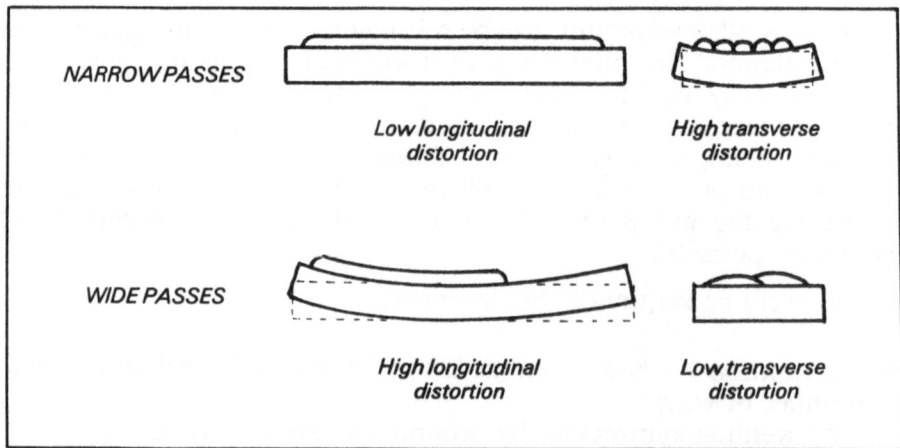

Fig. 8.20   Influence of bead size on distortions

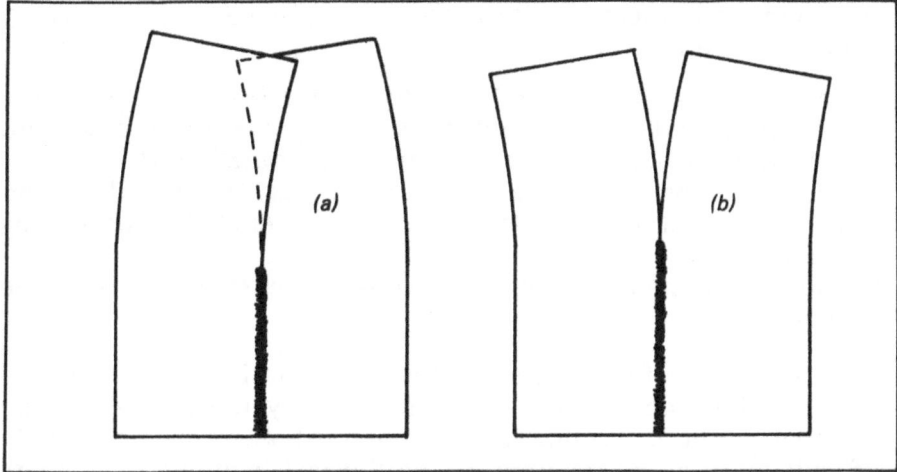

Fig. 8.21 Influence of welding speed on distortions due to transverse shrinkage

with reducing distortion and residual stress during the welding of light alloys by control of the heat flow in the metal. The idea was to use cryogenic liquids and auxiliary heat sources to produce expansions and contractions which would oppose those of welding.

Several of these techniques give welds with little or no distortion, but the only one which apparently gave repeatable results involved an arrangement which provided both cooling and heating (additional to that necessary for the welding).

Post-weld relief of residual stress
The object of this treatment is to cause the atomic structure of the metal to move, thus allowing the internal stresses either to relax or to adapt to the conditions of service. In a welded joint not subject to external stress it is possible to obtain total relaxation of the internal stress. If, however, the joint is stressed then there will also be internal stress.

The movement referred to is a displacement and a rearrangement of atoms. It is achieved either by means of mechanical shock, or by raising the metal temperature to increase atomic mobility. In both cases the the metal experiences local plastic deformation which leads to a reduction of the stress.

Mechanical methods. These include peening (hammering), vibration, and electrohydraulic or explosive methods. Peening is the method most often used since engineers are not unanimous regarding the efficiency and reliability of the other techniques.

Peening is the application of rapid regular blows, usually by a hand-held pneumatic hammer with a circular or rectangular head and rounded corners, or alternatively by a needle gun. Peening is mainly used for:

- The reduction or correction of distortions, especially angular ones (i.e. peening between weld passes to put surface bead into compression thus resisting the shrinkage contraction).
- The reduction of internal stress (note that peening thick and clamped parts only affects the surface layers of the metal being treated).
- The development of surface compressive stress (to improve fatigue life).

In fact the use of peening in welding should be kept to a minimum and should be considered as complementary rather than as the principal method. Peening is recommended where it is extremely difficult to carry out the appropriate thermal treatment (e.g. stress relief or normalising).

In the assembly of thick plates requiring a narrow joint preparation and on which preheating cannot be applied, then the peening of the intermediate passes will 'relieve' stresses minimising shrinkage.

When welding thick plates of materials which tend to be brittle when hot, the aim is, above all, to avoid the formation of cracks. Again peening can be an aid. In all cases, however, it is not necessary to systematically hammer all the passes.

Well-executed peening subjects the weld to compression such that a permanent plastic outflow of the surface material takes place. It is therefore usual to say that peening will be more effective on small rather than on large weld beads. Peening has little effect at depths greater than 5mm.

It should be noted that:

- It is necessary to avoid excessive peening (called overpeening), which no longer improves the stress situation but risks – either an excessive work hardening which can lead to cracks and the initiation – or an actual 'peeling' of the bead surface.
- Hammering is only effective when it is carried out with the metal in an elastic state, i.e. the temperature is below that where the metal can be readily plastically deformed. Hammering plastic material is ineffective and sometimes dangerous; it corresponds to bad forging. For certain steels (steels of high carbon content, Cr-Mo steels, etc.) it is preferable to hammer cold, thus avoiding the brittle temper temperature zone. Also carbon steels with more than 0.30% carbon and low alloyed steels should be given a subsequent thermal stress relief treatment or, better still, a normalisation treatment.

Vibratory stress relief. The technique aims to reduce residual stress by vibration, which is thought to act by improving atomic mobility. The equipment used consists of a vibration generator, often a spinning asymmetric mass. By varying the speed of rotation the frequency of vibration can be varied. The method is claimed to be effective when the frequency is close to the resonance frequency of the structure, the latter varying with the weight of the part. For most fabrications it is between 10 and 40 cycles per minute.

Stress relief by vibration may also be used to stabilise dimensions before machining. It is necessary to carry out a preliminary test to judge the effectiveness of the process; if the treatment is sufficient, the dimensional variations at the moment of machining are negligible. This procedure has the advantage of being both economical and relatively fast, but there is still considerable debate on its efficiency and reliability when applied to welded structures.

It is evident that vibration treatment should or can not be substituted for thermal treatment when the latter is applied for metallurgical reasons or when the thermal treatment is imposed by a construction code.

Thermal methods. Here, the necessary heat is obtained by furnace, oxy fuel burner, induction or resistance heating. Where a suitable furnace is available, this provides the preferable method, since the different parts of the fabrication are simultaneously brought to the same temperature. Heating by resistance or induction are also excellent methods. Control is more difficult with oxy fuel burners, especially where the treatment is applied locally on large welded assemblies.

Work hardening. Work hardening of a metal occurs following crystalline distortion resulting from cold mechanical working (e.g. cold rolling). Hardness, resistance to fracture and yield point all increase while ductility decreases, changes which can increase the risk of cracking.

Only an annealing treatment completely destroys the effects of cold rolling. Thermal stress relief treatment has a beneficial action in suppressing residual stresses but does not restore the crystalline structure.

Correction of distortions

Mechanical methods. The use of a press to cold straighten is limited by the possible risk of cracking or fracture. Hot straightening by mechanical methods reduces these risks.

Masubuchi discusses the electromagnetic hammer as a means of straightening. The principle of this arrangement is analogous to that of electromagnetic shaping: a coil carrying a variable current induces currents in the part which in turn tends to repel the part. Pulsing the coil current induces the hammering effect. The technique is particularly well adapted to materials which are good conductors of electricity, such as aluminium. With more resistive materials there is an energy loss resulting from resistance heating.

The electromagnetic hammer has been used to remove the distortions of welded tanks and bulkheads in the Saturn V rocket.

Thermal methods. The operation of flame shrinkage (or flame straightening), carried out with an oxyacetylene flame, allows the straightening of many types of components: plates, tubes, profiled sections, etc. The application detail varies according to the particular case, but the principle is always the same. Spot heating in the distorted (curved) region of the part causes expansion. The cold metal surrounding it opposes this expansion, the metal thickens and shortens, so that on cooling, the thermal contraction pulls the curve back into line. The phenomenon can be compared to that which gives rise to residual stresses.

The work of straightening a structure in sheet form, following welding, requires operators having extensive knowledge and experience. However, a Norwegian company has recently developed equipment giving greater control of the parameters which influence the treatment result. The concept is based on induction heating and electromagnets, i.e. clamping/positioning.

The induction element allows a concentration of heat on a defined zone of the sheet. The treatment is automatic and could therefore be robotised. The induction technique heats the surface skin resulting in an important difference in temperature between the two faces of the sheet. The result is shrinkage similar in concept to that of flame heating. Deep heating can be applied to the welded zones (for softening) and one can obtain equal temperatures on both faces of between 700 and 800°C. With this equipment, the heating temperature can be kept below 870–900°C, so that the structure of the metal is not modified. This also avoids 'burning' (oxidation) of the sheet, which is a fairly common defect when flame straightening, which remains visible even after final painting.

Chapter Nine

WELDING SAFETY

THIS chapter discusses the safety factors for three welding situations; namely, gas-shielded arc welding, laser welding, and automatic welding.

Gas-shielded arc welding

The hazards associated with gas-shielded welding processes which can threaten the health and safety of arc welders arise from:

- Radiation from the electric arc.
- Fumes, solid particles or gases emitted from the welding zone.
- Electric current.

Protection against radiation emitted by the arc

Nature of the radiation. The gas-shielded arc attains temperatures much higher than those observed in the use of covered electrodes, and this results in significant radiation from three parts of the electromagnetic spectrum: ultraviolet, visible and infrared. (The uv radiation emission being three or four times greater and other wavelengths as much as 30 times greater than from MMA. This radiation is emitted over a wide angle. In the absence of a slag the weld pool and its surroundings are 'mirror-like' with reflectivity which can be very high, especially with the light alloys. (In general fumes emitted absorb little radiation.)) Also the uv radiation increases in intensity with the current density, but its spectral distribution depends on the composition of the fused metal and the shielding gas.

Ultraviolet radiation of high intensity can burn organic matter and can disintegrate cotton material. Even brief exposure to uv rays can cause erythema of the skin, and their action on the human eye can be particularly painful, as detailed below.

The visible radiations can dazzle if sufficiently intense, and lead to visual fatigue. These radiations are particularly intense in the TIG and MIG processes for the reasons already given.

Infrared radiation, which is about 1.5 times more intense than in MMA welding, can over a long period cause cataracts and other injuries (see below).

Reflected rays of all wavelength ranges are dangerous and can be more insidious than direct rays, and people near the welding activity, even for a very short time, can be injured.

Although the tungsten electrodes used in TIG welding frequently contain (about 2%) thorium it has been established that any radioactivity resulting is so weak that it does not merit any special precautions.

Eye injuries and protection. Table 9.1 shows representative wavelengths and wavelength ranges pertaining to the infrared, visible and ultraviolet parts of the spectrum. Fig. 9.1 shows the response efficiency curve of the human eye to the visible wavelengths.

Infrared radiation is only harmful when averaging a prolonged exposure or in strong doses. The injuries caused are blepharitis, conjunctivitis, cataracts and burning of the sensory system. The thermal effects of ir rays are more rapidly felt than are the effects of the shorter wavelengths. Usually it is sufficient to consider the wavelengths from 8,000 to 15,000Å, but sometimes it is necessary to take account of higher wavelengths. As mentioned above, a strong irradiation of visible light from the whole visual field or a part of it, and also large variations in brightness (intensity), cause painful sensations: dazzling, flushing, ocular fatigue and general fatigue, headache, insomnia, etc.

Ultraviolet rays of 3,132–4,000Å wavelength do not appear to cause harm. However, the wavelengths less than 3,132Å cause acute effects on the anterior segment of the eye (conjunctiva, cornea, iris and lens) which after a latent period of six to eight hours becomes

Table 9.1   Wavelengths in the ir, visible and uv parts of the electromagnetic spectrum

Rays	Wavelengths (Å)	
	Representative	Range
Infrared	—	$7.500–3 \times 10^6$
Red	6,800	6,500–7,500
Orange	6,300	6,000–6,500
Yellow	5,900	5,700–6,000
Green	5,300	4,900–5,700
Blue	4,600	4,300–4,900
Violet	4,100	4,000–4,300
Ultraviolet	—	100–4,000

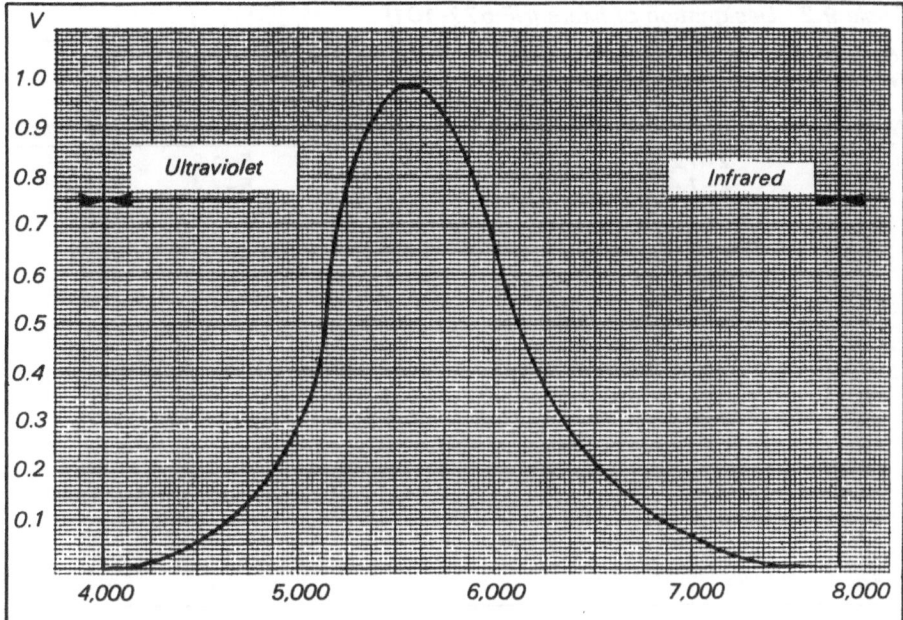

Fig. 9.1 Response efficiency curve of the human eye to visible wavelengths (NF 577-100)

extremely painful. With sensitive subjects the effect (known as 'arc-eye') can be induced following only a brief exposure or a weak dose. Repeated exposures during the same day have cumulative effects. There is never any tolerance. On the contrary, there may be an increased sensitisation which causes acute crises with lower doses or which manifest themselves by a chronic kerato-conjunctivitis with photophobia. The induced pain, watering of the eyes and photophobia are cared for effectively by treatment with anaesthetic and decongestant eyewashes.

There is therefore a need to protect eyes with screens which must fulfil certain criteria:

- Suppress or strongly reduce the ranges of infrared and ultraviolet wavelengths.
- Reduce the intensity (brightness) of visible rays, while ensuring sufficient light to distinguish the joint.

Comparison of Table 9.1 and Fig. 9.1 shows that green is the most suitable colour for the screens.

Tables 9.2 and 9.3 are extracts from AFNOR (French Standards Institute) standards showing standard filter numbers and for which processes they are recommended. (The equivalent British standards are BS1542:1982 and BS679:1959(1977).)

Table 9.2   Designation of filters (NF 577-101)

Welding filters NFS 77–104	Ultraviolet filters NFS 77–105		Infrared filters NFS 77–106	Daylight filters NFS 77–107	
Not numbered	Code No. 2	Code No. 3	Code No. 4	Code No. 5	Code No. 6
1.2	2–1.2	3–1.2	4–1.2	5–1.2	6–1.2
1.4	2–1.4	3–1.4	4–1.4	5–1.4	6–1.4
1.7		3–1.7	4–1.7	5–1.7	6–1.7
2		3–2	4–2	5–2	6–2
2.5		3–2.5	4–2.5	5–2.5	6–2.5
3		3–3	4–3	5–3	6–3
4		3–4	4–4	5–4	6–4
5		3–5	4–5		
6			4–6		
7			4–7		
8			4–8		
9			4–9		
10			4–10		
11					
12					
13					
14					
15					
16					

Body protection. No part of the body and particularly the face should be exposed to arc radiation if one wants to avoid burns analogous to strong sunburn.

Protection against fumes, gases and solid particles
Welding involves the emission of a certain quantity of fume, gas and particulate. It is important to prevent welding personnel from absorbing dangerous quantities of these substances. It is very difficult to give precise indications in this field, because different tests have shown that the quantities emitted vary considerably with the welding parameters (voltage, current, inclination of the weld torch, etc.). Also, for many substances, there is a lack of detailed knowledge relating to their chemical state and their real toxicity.

Carbon monoxide. In CO_2 and MAG welding, carbon dioxide, which serves as the protective shield, decomposes into carbon monoxide and oxygen:

$$2CO_2 \rightarrow 2CO + O_2$$

However, on the edges of the arc, recombination takes place so that there is only a minimum quantity of monoxide left free (Fig. 9.2). As the value of occupational exposure limits (abbreviated to OEL, Occupational Exposure Limits 1978 (Guidance Note EH 40/87 issued by the Health and Safety Executive and revised annually)), is 50ppm for carbon monoxide, it appears that its presence here is not dangerous. But since it may mix with other noxious products in the

Table 9.3 Levels of protection and recommended applications of filters for welding and gouging by electric arc (NF 577-104)

Process	Current intensity (A)																			
	10	15	20	30	40	60	80	100	125	150	175	200	225	250	275	300	350	400	450	500
Covered electrodes	▨																			
MIG on heavy metals		▨	▨	▨	▨	▨														
MIG on light alloys		▨	▨	▨	▨	▨	▨	▨												
TIG on all metals and alloys																▨	▨	▨		
MAG		▨	▨	▨																
Gauging by 'arc-air'		▨	▨	▨	▨	▨	▨	▨												

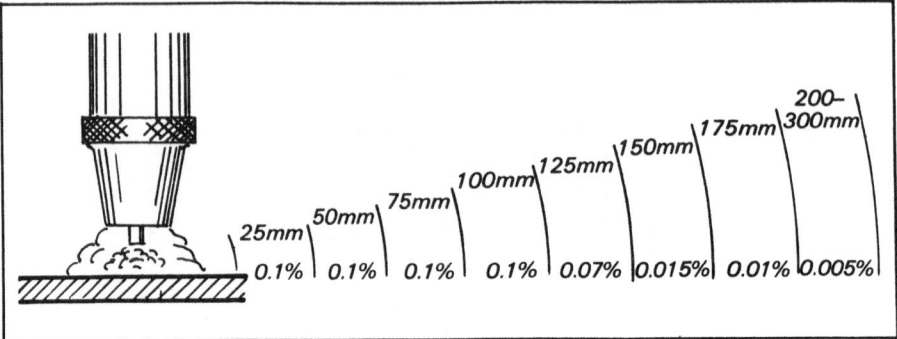

Fig. 9.2 Carbon monoxide levels in the vicinity of the arc in CO_2 welding

environment surrounding the welding zone, good ventilation remains indispensable. The CO build-up may be greater with automatic welding, where the welding production and hence output of gas is greater than with the manual or semi-automatic methods.

Ozone. Ultraviolet rays have the property of transforming oxygen (O_2) into ozone (O_3), which has an OEL of only 0.1ppm – a concentration which is often exceeded close to the arc. The symptoms of intoxication by ozone are depression and irritation of the eyes and nose. Ozone is, however, a fairly unstable gas and it scarcely persists close to the welder. Ozone emission can be high when welding aluminium (small quantities of nitrous oxide, or increased fume, can help reduce ozone levels).

Nitrous fumes. These fumes comprise the oxides of nitrogen, such as NO, NO_2 (OEL of 3ppm), N_2O_4, with the dioxide generally being preponderant. Their OEL values are seldom attained in present usage, but may be exceeded if mixtures of nitrogen and argon are used, for example, for the plasma cutting of stainless steels without providing adequate ventilation.

Phosgene. Certain chlorinated solvents such as trichloroethylene which are used in industry for degreasing are transformed by uv radiation from the electric arc into phosgene, a highly toxic gas (OEL of 0.1ppm). Dangerous concentrations can be present in the air in very low concentrations, too low to be detected by odour. It is therefore unacceptable to clean components to be welded with these chlorinated solvents.

Fumes. In TIG welding, if the operating mode is incorrect, it can produce an accidental vaporisation of the tungsten electrode, easily recognised by the appearance of whitish fumes which are very toxic (OEL of 1mg/m³ per 8 hour exposure). Metal vapours are also produced in MIG welding. They form into clouds of very fine, largely metallic, dust which act on the lungs and the stomach and can produce bouts of fever. The metals arise from electrodes (the wire or their metal or flux core) or from coatings (e.g. zinc, OEL of $5mg/m^3$ per 8 hour exposure) on the sheets being welded. Vaporisation from the electrode occurs more readily when welding currents or current density is high. It is often the alloying elements which vaporise, e.g. manganese (OEL of $5mg/m^3$) from Si-Mn steel or magnesium from aluminium alloys. Copper vaporisation (OEL of $0.2mg/m^3$ as fume, $1mg/m^3$ as dust and mists) can arise from the protective coating on steel electrode wires, or when welding copper and its alloys.

A conference held in Copenhagen (February 1985) and organised jointly by the EEC, the International Centre for Cancer Research, the Danish Welding Institute and the European Bureau of the World Health Organisation (WHO), discussed the effects of all particles and gases emanating from welding, especially from welding consumables high in chromium and nickel. It was recognised that chromium VI was highly toxic and carcinogenic as reflected in an OEL of $0.5mg/m^3$. It is not known whether the fumes/particulates emitted from welding stainless steels is toxic to this degree.

However, although certain experts think that the presence of chromium and nickel in metallic fumes emitted during welding of stainless steels can be carcinogenic, this risk is not scientifically proven and it is for this reason that the WHO decided to conduct, together with the whole international scientific community, an extensive epidemiological study.

Argon and carbon dioxide. These gases are not toxic, but carbon dioxide (OEL of 5,000ppm) accelerates respiration and can thus increase the dosage of other noxious fumes inhaled. In confined areas they can displace oxygen leaving an environment which could lead to asphyxiation.

To protect against gases and fumes it is necessary to provide ventilation close to the arc. However, care must be taken that the ventilation does not create air currents which prevent the shielding gas from performing its task. Generally, the maximum allowable air flow rate is about 30m/min.

There are numerous arrangements for the capture of fumes directly at the welding torch. These techniques may be effective for manual welding, but they are often less well adapted to robotised welding simply because of additional encumbrance due to the ventilation tube both at the torch, and along the welding conduit length. However, these arrangements can nevertheless prove to be useful when the robot is equipped for self-shielded flux-cored welding which can emit copious fumes. The lack of need for shielding gas results in less voluminous torches which can more easily support the extraction arrangement.

Welding with robots or automatic machines permits hooding and total area extraction of fumes. The evacuated air is either expelled from the building or recirculated in the workshop after passage through an appropriate filter system.

Fig. 9.3 suggests guide values for the number of local air changes per hour in relation to the average level of welding current used. They may be helpful in selecting the choice of ventilation method.

It should be noted that the automatic welding processes tend to emit less fumes per kilogram of deposited metal than do the manual metal arc processes (Table 9.4).

Other safety measures
It is important to consider a number of other hazards and to offer protection against them; for example, against the danger of electrocution, the possible splashing of molten metal and from the explosive release of compressed gases. Appropriate rules can be summarised as follows:

- Ensure that equipment is correctly installed, and that the electrical equipment is earthed.
- Avoid impact shock to gas cylinders, which should be fixed to a stable support. Do not expose gas cylinders to temperatures above 55°C. Ensure that they are subject to regular maintenance.
- Take appropriate measures (i.e. correct working practices, protective clothing and screens, etc.) to protect against possible splashing of molten metal (spatter) and from hot workpieces.

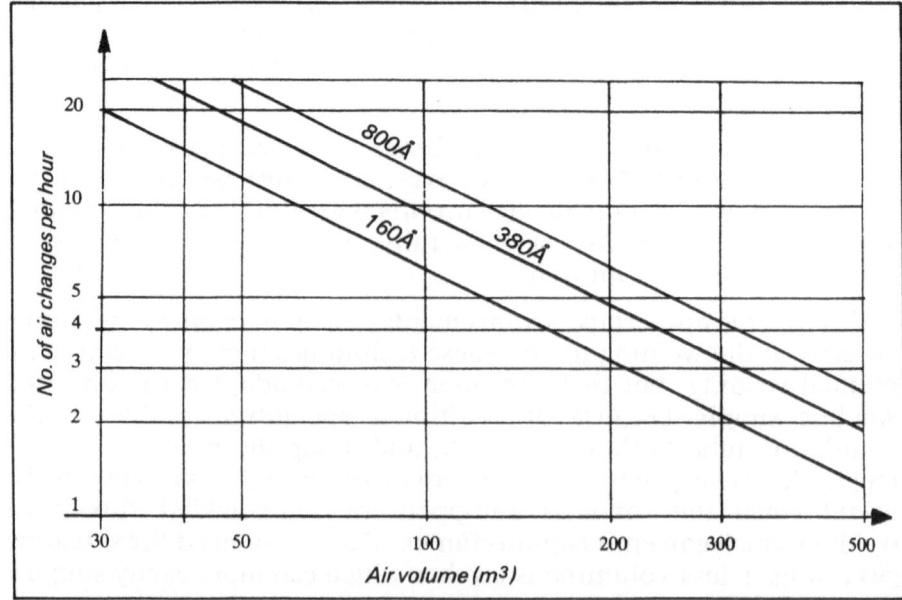

Fig. 9.3 *Recommended air changes per hour in welding workshops*

Table 9.4 Comparison of levels of welding fume emitted by various processes

Process	Rate of deposition (kg/hour)	Quantity of fumes released (g/min)	Quantity of fumes per kg of metal deposited (g)
Covered electrode	2	0.7	21
High-yield electrode	5	1.3	16
Solid wire/gas	1.5	0.5	8.5
Covered wire/gas	7	1.4	12
Covered wire without gas	7	2.2	18

Workplaces for gas-shielded welding

To minimise the reflection of arc radiations, particularly of uv radiation, the walls of the workplace should present a gritty or matt surface (matt and not gloss paints should be used).

The welding bays or stations should be surrounded by solid screens, opaque (i.e. canvas) curtains, or translucent uv absorbing plastic curtains to protect the casual observer from radiation. If one is welding large assemblies, where it is not possible to surround the welding area with curtains, it may become necessary to create corridors protected from radiation or to ensure all personnel wear protective glasses. (Even clear glass significantly reduces the risks of eye damage from uv radiation.)

Laser welding

Lasers generating infrared radiation are often used in welding. They constitute a danger, particularly for the lens of the eye, as the radiation is invisible and the natural protective mechanisms (e.g. blinking) do not operate to prevent exposure. For CO_2 lasers with emission in the ir range, Plexiglas protection normally suffices. However, as laser power increases (e.g. 10–25kW) more sophisticated protection devices are required. Ruby (YAG) lasers, whose emission wavelength is about six times smaller (1.06μm), present greater dangers, and indeed the complete hooding of the machine or its isolation in a separate workshop is necessary. (The safety requirements are more fully described in national safety documents such as BS4803:1983, Radiation Safety of Laser Products and Systems; Part 1 – General; Part 2 – Manufacturing Requirements for Laser Products; Part 3 – Guidance for Users.)

Automatic welding

We have just discussed the problems inherent in fusion welding and the following risks are easy to identify:

- Risk due to electric currents.
- Risk due to radiation.
- Risk due to gas, vapours, and fumes.
- Risk due to splashing of molten metal.
- Risk due to contact with hot metal.

Automatic welding (by machine or robot) and the associated loading/unloading of parts into and out of the welding cell removes the operator from these hazards. However, the use of robots can introduce new problems, particularly those associated with the fast, and often to the casual observer, unexpected movement. These risks are to be discussed in the other volumes of this series.

BIBLIOGRAPHY/ FURTHER READING/ REFERENCES

Manuals

Précis de Sougdage, Brasage et Techniques Connexes. R. Le Gouic. Editions Eyrolles.

Guide de Soudage (5 volumes). Syndicat National des Fabricants de Matériels de Soudage. Editions Gead.

Conception des Assemblages Soudés. J. G. Hicks. Editions Eyrolles.

Soudage (Cours du Conservatoire National des Arts et Métiers). H. Gerbeaux. Soudure Autogène.

Les Procédés de Soudage. P. T. Houldcroft. Editions Dunod.

Joints Soudés – Contrôle, Métallurgie, Résistance. A. Vallini. Editions Dunod.

Guide de Soudage Semi-automatique et Automatique. Nertalic MIG-MAG-Safdual Safuni.

Guide l'Utilisateur des Procédés de Soudage sous Atmosphère Neutre (TIG). Nertal. Nertal-Point-Miniplasma. Soudure Autogène.

Le Soudage avec Fil Fourré (FabCO). Hobart Brothers AG, Amsterdam. La Soudure Exotherme.

Le Soudage Électrique par Fusion. Sciaky.

Journals

L'Usine Nouvelle

Détection et analyse non destructives et constraintes dans les matériaux. J. Antoine (December 1982).

Soudage laser. D. Coué (February 1983).

Le laser se met en six. C. Desforges (October 1977).

Soudage à l'arc: Quatre gaz valent mieux qu'un. D. Coué and J. Nenin (March 1983).

Souder

Quels sont les facteurs qui déterminent la forme des soudures MIG/MAG? B. Haas and H. U. de Pomaska (No. 151).

Serrurerie – Constructions Metalliques

Le soudage automatique et semi-automatique avec fil fourré sous gaz. J. R. Cossmann and H. B. Gary (No. 199).

Welding and Metal Fabrication

Analyse du transfert de métal en soudage MIG. J. Ma and R. L. Apps (April 1983).

Industrie Anzeiger

Possibilités et limites dans l'application du soudage TIG avec arc pulsé et sources transistorisées. L. Dorn and P. Jahn (No. 23, 1983).

Soudage TIG par impulsions au moyen de générateurs de courant à transistors. L. Dorn and P. Jahn (No. 41, 1983).

DVS

Développements récents quant au transfert de métal précis par le procédé GMA pulsé. R. R. Wright, P. F. Rogers and C. J. Allum.

Générateur de courant de soudage électroniques avec interface pour robots industriels et commande par automate. P. Puschner.

Soudage et Techniques Connexes

Impact de la recherche sur le développement du soudage à l'arc. A. Smith, Welding Institute, UK.

Programme National de Recherche en Soudage. Comité Scientific et Technique de la Société des Ingénieurs Soudeurs, France.

Evolution et tendances du soudage automatique à l'arc. P. Berthet and J. Panisset, Institut de Soudure, France.

Généralités sur le soudage avec fil fourrés en France. M. Mahe, La Soudure Exotherme, France.

L'arc électrique en construction soudée, état actuel et perspectives. C. Messager, CTAS: Air liquide/SAF, France.

Examen de nouvelles conceptions de source de courant pour les procédés de soudage à l'arc. Doc. 115/IIW-701-82. Commission XII de l'IIS, France.

Choix du procédé de soudage. D. Van Der Torre, Metaalinstituut, Netherlands.

Soudage des métaux par laser YAG. P. Aubert, C. Decailloz, J. Y. Gascoin and H. Staegr, CEA, France.

Le laser CO_2. H. Herbrich, W. Scheuermann and D. Weskott, Messer Griesheim, West Germany.

Travaux récents de IS dans le cadre des applications laser. C. Kluzinski and J. Rosenzweig, Institut de Soudure, France.

Soudage laser de structures sandwich métalliques du type Norsial. J. Haroutel, SNIAS, France.

Soudage par faisceau laser: Etat actuel de la question et perspectives d'avenir. C. Kluzinski, Institut de Soudure, France.

Caractéristiques de l'arc TIG – Etude des paramètres définissant le bain de fusion. J. Binard, Frumatome, Centre de Soudage, France.

Ecoulement dans les bains métalliques en procédé de soudage TIG. Y. Fautrelle, Madylam, France.

Etude expérimentale des paramètres influençant le bain de fusion (TIG). J. Binard and A. Chabenat, Frumatome, Centre de Soudage, France.

Etude des plasmas d'arc du point de vue du soudage. M. Evrard, Institut de Soudure and B. Blanchet, SNECMA, France.

Le soudage plasma, P. Demars.

Machine Moderne
Les préparations avant soudage. A. Devann (December 1976).

Technical Bulletins

Le laser de 10kw, AHPL-10 et le travail des métaux. Sciaky Technical Notebooks.

Procédés de soudage à l'arc plasma. M. Aloïs and H. Wagenleitner. Scheron.

La technique du soudage automatique à l'arc, Vol. 48. P. Nobbe.

Application pratique du procédé de soudage TIG aux aciers, aux métaux non ferreux taux alliages de nickel, Vol. 48. A. Schmid. Brown Boveri.

Conferences

Maîtrise des déformations dues au soudage. M. de Bony, 10 June 1976, Centre Technique de l'Aluminium, France.

Conférences sur la Technologie de Soudure au Japon, 22–23 June 1983, London.

Robotique de la soudure à l'arc au Japon. M. Kiyohara, OTC, Japan.

Progrès récents des matériels de soudage à l'arc et leurs applications. M. Hideyuki Yamamoto, OTC, Japan.

Contrôle multiparamètre des tubes par courants de Foucault. M. Pigeon, CEA-DT.SECS.STA, France.

La Mécanique à l'Heure des Lasers de Puissance, CETIM de SENLIS, March 1985.

Les lasers de puissance. M. M. C. Kluzinski, A. M. Lemaire, D. Marchand and P. Piednoir, Institut de Soudure, France.

Soudage et Traitements Thermiques, Journées de Conférences, December 1978, Marseille. Société Française de Metallurgie et la Société des Ingénieurs Soudeurs, France.

Traitements thermiques avant, pendant et après soudage. H. Granjon, Institut de Soudure, France.

Moyens d'application des conditionnements et traitements thermiques aux constructions soudées. P. Berthet, Institut de Soudure, France.

Soudure des alliages Al-Zn-Mg. M. Bousseau, DRET-ETCA-Arcueil, France.

L'utilisation des gaz dans le soudage. M. Robic, L'Air Liquide, France.

Commercial Documentation

The following companies supplied the authors with commercial documentation:

Arcos
Ardrox
Automatisation Internationale

AWP (GEC Industrial Controls Ltd)
BOC – Arc Equipment
BOC Industrial Power Beams
CLOOS
Commercy Soudure
Cyclomatic
Elma Technik
ESAB
Fronius
Hitachi
Hobart (La Soudure Exotherme)
Kobe Steel Ltd
Laboratoires de Marcoussis–CGE
L'air Liquide
Lincoln Electric
Linde
Mansfeld
Messer Griesheim
Murex, Etarc, Sarazin Soudure
Oerlikon (Secheron Soudure)
Osaka Transformer Co.
Philips Welding (FILARC)
Polysoude
SAF
Spectra- Physics
Union Carbide (L TEC)

References

Aichele G. and Smith A. A. *Mag-Schweissen*, Deutscher Verband fur Schweisstechnik, West Germany.

Allen B. C. and Kinergy W.-D. Surface tension and contact angles in some liquid metal solid ceramic systems at elevated temperature. *Trans. Met. Soc.*, AIME, 215: 32–36, 1959.

Allen B. C. The surface tension of liquid metals. In, *Liquid Metals – Chemistry and Physics*, pp. 161–212. S. Z. Beer, New York, 1972.

Amin M. and Lucas W. *Welding Institute Research Bulletin*, 15 (3): 74–78, 1974.

Amin M. Prediction of square wave pulse current parameters for control of metal transfer in MIG welding. Welding Institute Members Report 83/1978/P.

Amin M. *Metal Construction*, 13 (6): 349–353, 1981.

Amin M. *Welding Institute Research Report*, 83/1978/1, 1978.

Amin M. *Welding Institute Research Bulletin*, 271–275, 1981.

Ancelot D. and Landry J. Soudage longitudinal des tubes à grande vitesse par procédé automatique sous flux. *Soudage et Techniques Connexes*, November/December, 1981.

Anderson D. Streaming due to a thermal surface tension gradient. *Weld. Res. Abroad*, January: 55–64, 1974.

Anderson N. E., Hights B. and Greene W. J. *United States Patent Office*, Brevet No. 3 071 680.

Ando K. *et al.* Mechanism of formation of pencil-point-like wire tip in MIG arc welding. *IIW Doc.*, 212-156-68.

Andrews J. G. and Craine R. E. Fluid flow a hemisphen induced by a distributed source of current. *J. Fluid Mech.*, 84: 281–290, 1978.

Araya T. *et al.* Transistor-controlled pulse MIG welding of aluminium alloys, IIW-XII-C19-81.

Asuka K., Nishayama N. and Tsuboi, T. *Metal Construction*, 13 (9): 570-574, 1981.

Baba Y., Fukui T., Takashima A. and Terai, S. *Newly developed Al-Zn-Mg alloys containing less magnesium and their application to transport industry.* In, *6ᵉ Congrès International des Métaux Légers*, Leoben, Vienne, 1975. Alu. Verlag GmbH, West Germany.

Bailey N. *Welding Journal Research Supplement*, 67 (4): 169s-177s, 1972.

Bastien B., Roques C., Dollet J., Rapenne H. and Pitaud J. *Soud. Tech. Conn.*, 16 (7-8): 257, 1962.

Baysinger F. R. Distortion control of welded aluminium structures. *Weld. Int. Suppl.*, 51 (12): 833–841, 1972.

Becker W. Apport de chaleur contrôlé pour le soudage TIG. *Der Praktiker*, 24 (10): 201-204, 1972.

Bernard G. A view point on the weldability of C-Mn and microalloyed steels: Microalloying 75. *Rapport IRSID*, 245, 1975.

Bernard G., Faure F. and Gauthier G. Ténacité des zones affectées par la chaleur du soudage des aciers C-Mn et microalliés. *Rapport IRSID*, 1976.

Binard J. Etude bibliographique en vue de préparer un programme approfondi de recherche sur le comportement de l'arc TIG et du bain de fusion. June 1982, Action concertée: CPS, Décision d'aide No. 79-7-1516, Rapport d'étude CATS le Creusot, No. 361.

Blodgett O. W. Distortion: How metal properties affect it. *Weld. Engr. USA*, 57 (2): 41-46, 1972.

Blommel G. Le nouvel appareil de commande WIG (TIG) type STG-251 pour le soudage avec arc pulsé. *Oerlikon-Mitteilungen*, 31 (65-66): 17-19, 1971.

Bonizewski T. *British Welding Journal*, 5 (1): 225-229, 1969.

Bonnet C. Relations structure – résilience dans les soudures d'acier doux et faiblement alliés brutes de solidification. *Soudage et Techniques Connexes*, 34 (7-8): 209-228, 1980.

Boughton P. Le procédé de soudage TIG avec arc pulsé, 2ᵉ partie. Possibilités d'application. Research Member Seminar, London, December 1972. Welding Institute, Abington Reports and Papers, 5-8.

Bradstreet B. J. Effect of surface tension and metal flow on weld bead formation. *Weld. Res. Suppl.*, July: 314-322, 1968.

Buchinskii V. N. and Voropai N. M. Features of the pulsed-arc welding of steels using a mixture of argon and CO_2. *Avt. Svarka*, 3, 1978.

Burkdekin F. M. *Brit. Welding J.*, February: 81, 1967.

Capel L. Aluminium welding practice. *Brit. Weld.*, 8 (5): 245-257, 1961.

Cebulak W. S. and Truax D. J. Program to develop high strength aluminium powder metallurgy products. Final Report, 29 September 1972, US Army Frankford Arsenal Contract DAAA 25.70. CO358.

Cervay R. R. Engineering design data for aluminium alloys 7475 T761 and T61 condition. Technical Report AFML-TR-72-173, September 1972.

Cheviet A. and Murry G. Etude du traitement de relaxation d'aciers de type E36. Résultats publiés dans le rapport CEDA Eur 5185 f, December 1974.

Cheviet A. and Murry G. Les traitements de détensionnement des constructions soudées, Rapport IRSID RE 75, June 1971.

Chevigny R., Develay R., Guilhaudis A. and Petrequin J. Les alliages aluminium-magnésium: propriétés mécaniques, soudabilité, résistance à la corrosion, état structural. *Alluminio Nuova Metallurgia*, 10: 507-529, 1965.

Cline C. L. Weld shrinkage and control distortion in aluminium butt welds. *Weld Int. Suppl.*, 523S-528S, 1965.

Coe F. R. *Welding Steels without Hydrogen Cracking*. Welding Institute, Doc. II-682-73 de l'IIS.

Coe, F. R. *Le Soudage dans le Monde*, 7 (1): 16-26, 1969.

Coherent Inc. *Lasers*. McGraw-Hill Book Company, 1980.

Connor L. P., Rathbone A. M. and Gross J. H. Effects of composition of the heat affected zone toughness of constructional alloy steels. *Welding J.*, (5): 217s-234s, 1967.

Conserva M., Di Russo E. and Gatto F. A new thermomechanical treatment for Al-Zn-Mg type alloys. *Alluminio*, 37: 441-445, 1968.

Cowdery D. L'opération de soudage TIG arc pulsé 1re partie – étude et réalisation de l'équipement. Research Member Seminar, London, December 1972. Welding Institute, Abington. Reports and Papers, 1-4.

Craine R. E. and Weatherill N. P. Fluid flow in a hemispherical container induced by a distributed source of current and a superimposed uniform magnetic field. *J. Fluid Mech.*, 99: 1-12, 1980.

Craine, R. E. and Andrews, J. G. The shape of the fusion boundary in an electromagnetically stirred weld pool. *Iutam Symposium on MHD and its Metallurgical Applications*, Cambridge, 1982.

De Keyser. Nouveaux types de sources de courant pour le soudage à l'arc. Doc. XII-F-204-79 de l'IIS.

Desre P. J. and Joud J. C. Surface tension temperature coefficient of liquid alloys and definition of a zero Marangoni number alloy for crystallisation experiments in microgravity environment. *Acta Astronautica*, 8: 407-415, 1981.

Develay R. Alliages Al-Zn-Mg à moyenne résistance et à haute fiabilité. *Revue de l'Aluminium*, 349 (January): 86-99, 1967.

Develay R. Les alliages d'aluminium corroyés à moyenne et haute résistance. Propriétés métallurgiques. Critères de choix. *Bulletin du Cercle d'Etudes des Métaux*, 8 (6): 277-332, 1976.

Develay R. Les aliages d'aluminium à haute résistance pour l'industrie aérospatiale. *Revue de l'Aluminium*, February: 156-184, 1972.

Di Russo E. and Buratti M. Carrateristiche di due nuove leghe d'alluminio da larorazione. Plastica: Zergal 3, Zergal 4. *Alluminio*, 1: 31-41, 1974.

Di Russo E., Conserva M., Buratti M. and Gatto F. A new thermo-mechanical procedure for improving the ductility and toughness of Al-Zn-Mg-Cu alloys in the transverse direction. *Materials Science and Engineering*, 14: 23-36, 1974.

Di Russo E. Indagini sperimentali su leghe complesse Al-Zn-Mg tenori controllati di chromo, zirconio argeno. *Alluminio*, 10: 505-519, 1964.

Di Russo E. and Signoretti S. Improvement of the properties of high strength Al-Zn-Mg-Cu alloys by thermomechanical procedures. *Agard. Report*, 610, 1973.

Di Russo E., Conserva M., Gatto F. and Markus U. Thermomechanical treatments on high strength Al-Zn-Mg (Cu) alloys. *Metallurgical Transactions*, 4 (April): 1133-1144, 1973.

Dorn L. and Jahn P. L'influence de la superposition d'impulsions lors du soudage TIG à source de courant transistorisée. *Schweisstechnik* (Zurich), 68 (5): 95-100, 1978.

Drews P. and Puschner, P. La source d'énergie de soudage transistorisée. *Schweissen und Schneiden*, 26 (2): 54-56, 1974.

Essers, W. G., Willems G. A. M. and Buelens J. J. C. Pays-Bas: Soudage plasma MIG d'aluminium. *Colloque sue l'Aluminium et ses Alliages en Construction Soudée*. Institut International de la Soudure, Porto, September 1981.

Eustathopoulos N. and Joud J. C. *Interfacial Tension and Absorption of Metallic Systems. Current Topics in Materials Science*, Vol. 4, Ed. by D. Kaldis. North-Holland, 1980.

Evans G. M. and Baach H. Doc. II-726-74 de l'IIS (XII-B-175-74).

Fihey J. L. and Simoneau R. Weld penetration variation in GTA welding of some 304 L stainless steels. *Welding Technology for Energy Applications: Conference AWS*, 16-19 May 1982.

Forch K., Forch U. and Piehl K. H. Interprétation par la métallurgie physique de la variation de la résilience de la zone affectée thermiquement lors du soudage d'aciers de construction. *Stahl und Eisen*, 98 (13): 641-651, 1978.

Gaillard F. Influence des traitements de mise en solution sur la structure et la résistance à la corrosion feuilletante d'une tôle en alliage AZ5G. *Rapport STCAN*, 2869, 1971.

Gatto F., Di Russo E., Conserva M. and Buratti, M. Su un trattamento termomecanico per migliorare la duttilita, la tenacita e la resistenza alla tensocorrosione delle leghe Al-Zn-Mg-Cu ad elevata resistenza meccanica. *La Metallurgica Italiana*, 11, 1974.

Gentilicore V. J., Pense, A. W. and Stout R. D. Fracture toughness of pressure vessel steels weldments. *Weld. J.*, 49 (8): 341s-353s, 1970.

Gilde W. Transverse weld shrinkage. *Weld. Int.*, 37 (2): 485, 1958.

Gosman A. D., Pun W. M., Runchal A. K., Spalding D. B. and Woflshtein M. *Heat and Mass Transfer in Recirculating Flows.* Academic Press, 1969.

Gott C., Clover F. R. and Rudy J. F. Metal movement and mismatch in aluminium welds. *Weld Int. Suppl.*, 47 (8): 337S-344S, 1968.

Granjon H. and Debiez S. Evaluation, par la méthode du double implant, du risque de fissuration lors du traitement de relaxation des ensembles soudés sur acier. *Rev. Mét.*, 70 (12): 1033, 1973.

Gray C. N. *et al. Weld Pool Chemistry and Metallurgy, International Conference*, London, 1980.

Grist F. J. and Armstrong, F. W. A new ac constant potential power source for heavy plate, deep groove welding. *Welding Journal*, June 1980.

Gulvin T. F., Scott D., Haddrill D. M. and Glen J. The effect of modern fabrication techniques on the properties of steel. *J. of the West of Scotland Iron and Steel Institute*, 80: 144-171, 1972/73.

Hartmann F. *Les Lasers*. PUF, 1974.

Haure, J. and Bocquet P. Fissuration sous les revêtements inoxydables des pièces pour cuve sous pression. *Rapport final CEDA*, Convention 6210-73-3 1303.

Heiple C. R. and Roper J. R. Mechanism for minor element effect on GTA fusion zone geometry. *Welding Journal, Weld Res. Sup.*, April 1982.

Heiple C. R. and Roper J. R. Effect of selenium on GTAW fusion zone geometry. *Weld Res. Suppl.*, August: 143-145, 1981.

Heiple *et al.* Surface active element effects on the shape of GTA, laser, and electron beam welds. *Welding Journal – Weld. Res. Sup.*, March 1982.

Heiro H. and North T. H. The influence of welding parameters on droplet temperature during pulsed arc welding. *Welding and Metal Fabrication*, September 1976.

Hondros E. D. *Physico-chemical Measurements in Metal Research*, 4 (2): 293, 1970.

Hunsicker H. Y., Staley J. T. and Brown R. H. Stress corrosion resistance of high strength Al-Zn-Mg-Cu alloys with and without silver additions. *Metallurgical Transactions*, 3 (January): 201-220, 1971.

Jones R. E. Mechanical properties of 7049 T73 and 7049 T76 aluminium alloys extrusions at several temperatures. *Rapport AFML-TR-72-2*, February 1972.

Joud J. C., Desre P., Pichard C. and Poyet P. Compte rendu d'êtude DGRST, September 1981.

Joy G. D. and Sage A. M. *Stress Relief Heat Treatments of Vanadium Steels*. Highweld Publications, 1972.

Kaufman J. G. Design of Aluminium Alloys for High Fatigue Strength. 40th Meeting of the Structures and Materials Panel, Agard, 15 April 1975, Bruxelles, Rapport Agard Cp 185, 2.1-2.26.

Keene B. J., Mills K. C. and Robinson J. L. The effect of surface tension on variable penetration behaviour during the mechanised TIG-welding of austenitic stainless steel. *Int. Conf. on the Effects of Residual, Impurity and Micro-alloying Elements on Weldability and Weld Properties*, London, 15-17 November 1983. The Welding Institute.

Koch H. Solidification contrôlée des bains de fusion. *Soudage et Coupage*, 27 (1): 29-30, 1975.

Koike Y. *et al.* Doc. XII-B-89-71 de l'IIS.

Konigshofer T. Les sources de courant de soudage en question. *Schweissen und Schneiden*, 25 (8): 303-305, 1973.

Krantz B. M. and Coppolecchia V. D. The effects of pulsed gas metal-arc welding parameters on weld cooling rates. *Welding Journal*, November 1971.

Krause R. D. and Schurmeyer R. Le soudage TIG avec courants pulsatoires sur les aciers fortement alliés. *Rapport D VS 25*, 37-42 Deutscher Verlag fur Schweisstechnik, Dusseldorf, 1972.

Lauprecht W., Emrich P. and Speth W. Interchangeability of temperature and time in stress relieving of steel joints. *Weld Research Abroad*, April: 43-48, 1972.

Lawrence F. V. *Welding Journal*, May: 212s, 1973.

Lenivkin V. A. *et al.* The determination of regions of controllable metal transfer in pulsed arc welding with a consumable electrode. *Svar. Proiz*, 1, 1976.

Leymonie C. Etude de la relaxation anisotherme de quatre aciers. *Congrès sur les Aciers Résistants à Haute Température*, Dusseldorf, 1972.

Lucas W. Utilisation de sources de courant transistorisées en soudage à l'arc. Doc. XII-F-226-80 de l'IIS.

Lucas W. The application of the synergic pulsed process and the influence of the weave pattern in mechanised MIG-welding. DVS 68.

Lucas W., Street J. A. and Watkins P. V. C. *The Welding Institute Members Report*, 66-75, 1975.

Luhau J. V. and Summerston T. J. Development of 7 042 T73 high strength stress corrosion resistant aluminium alloy forgings. *Metals Engineering Quarterly*, November: 35-42, 1970.

Lyle J. P. and Cebulak W. S. Powder metallurgy approach for control of microstructure and properties in high strength aluminium alloys. *Metallurgical Transactions A*, 6A (April): 685-699, 1975.

Lyle J. P. and Cebulak W. S. Properties of high strength aluminium P/M products. *Metals Engineering Quarterly*, February: 52-63, 1974.

Ma, Jilong and Apps R. L. MIG study yields important practical results. *Welding and Metal Fabrication*, December 1980.

Mahn G. T. and Rosenfeld A. R. Metallurgical factors affecting fracture toughness of aluminium alloy. *Metallurgical Transactions A*, 6A (April): 653-668, 1975.

Marchand D. and Kluzinski C. Application des lasers au micro-soudage. *Opto Electronique*, 21, July/August 1984.

Markworth, M. Kupferhaltige Al-Zn-Mg Knetlegierungen and Entwicklungstendenzen. *Aluminium*, 48: 724-732, 1972.

Martin G. and Bosseau M. Tenue en fatigue des assemblages en 7 020 soudés bout à bout. *Rapport ETCA-1*, 78.

Masubuchi K. Residential stresses and distortion in welded aluminium structures and their effects on service performance. *WRC Bull. USA*, 174: 1-30, 1972.

Matsunawa A. and Ohjit. Role of surface tension in fusion welding. *Transactions of JWRI*, 11 (2): 145-154, 1982.

Mazzolani F. M. Les imperfections structurales dans les assemblages soudés en aluminium. *Revue de l'Aluminium*, 431 (July/August): 420, 1974.

McDowel R. O., Hinkeldey P. L. and Schimmel H. W. Evaluation of premium strength stress corrosion 7000 series aluminium alloy die forgings. *Metals Engineering Quarterly*, February: 45-46, 1970.

McMillan J. C. and Hyatt M. V. Development of high strength aluminium alloys improved stress corrosion resistance. *Rapport AFMC-TR-68-148*, June 1968.

Moulin J., Adenis D. and Develay R. Interprétation des anomalies dilatométriques de l'alliage AZ5G par microscopie électronique. *Mémoires Scientifiques de la Revue de Métallurgie*, LXIV, No. 11, 1967.

Muller W., Ingenbrand H. D. and Borutzki U. Soudage TIG avec arc pulsé. *ZIS-Mitteilungen* (Halle), 11 (7): 1077-1078, 1969.

Murphy S. and Woodhead J. H. An investigation of the validity of certain tempering parameters. *Met. Trans.*, 3 (3): 727-735, 1972.

Myers J. Effects of deoxidants and impurities on 'simulated' stress relief cracking of 1/2 Cr 1/2 Mo 1/4 V steel. *Int. Conf. on Welding Research Related to Power Plant*, September 1972.

Needham J. C. Procédé TIG avec arc pulsé – Une introduction à la méthode. *Research Member Seminar*, London, 7 December 1972. Welding Institute, Abington.

Needham J. C. and Boughton P. *Welding Institute Research Bulletin*, 13 (2): 37-42, 1972.

Needham J. C. and Carter A. W. *British Welding Journal*, 14 (10): 533-549, 1967.

Needham J. C. and Smith A. A. Welding Institute Contract Report C 259/1/69 for Republic Steel.

Needham J. C. and Smith A. A. *British Welding Journal*, 3 (2): 45-51, 1955.

Needham J. C. and Boughton P. Une approche à la rationalisation des sources de courant en soudage à l'arc. Doc. XII-F-190-78 de l'IIS.

Needham J. C. and Carter A. W. *British Welding Journal*, 17 (5): 229-244, 1965.

Needham J. C. Sinusoidal and square wave pulsed currents for MIG welding. Institut International de la Soudure, Doc. XII F. 112-71.

Needham J. C. *Welding Institute Research Bulletin*, 21 (2): 47-52, 1980.

Needham J. C. Synergic pulse MIG welding. *Welding Institute Research Bulletin*, 18 (September), 1977.

Needham J. C. Pulse controlled consumable electrode welding arcs. *British Welding Journal*, April 1965.

Nicholson S. and Brook J. C. Review of codes with reference to heat treatments. Doc. IIS/IIW X 680-72.

Nishiguchi K. Evaluation des sources de courant continu utilisées en soudage sous protection gazeuse, particulièrement pour les procédés MIG/MAG. Doc. XII-F185-77 de l'IIS.

Olinger I. P. Mise au point d'une source de courant de soudage transistorisée. *Schweisstechnik* (Vienne), 24 (10): 139-42, 1970.

Oreper G. M. and Szekely J. On electromagnetically and buoyancy driven flow in weld pools. *J. Fluid Mech.*, 1983.

Oreper G. M., Eagar J. W. and Szekely J. Convection in arc weld pools. *Welding Journal*, 62: 307, 1983.

Orszag, A. and Hepner, G. *Les Lasers et Leurs Applications*, Masson, 1980.

Pachleitner P. Stand und Trend der Entwicklung von AlZnMg Legierungen, *6ᵉ Congrès International des Métaux Légers*, Leoben, Vienne. Aluminium Verlag GmbH, Dusseldorf, pp. 76-78, 1975.

Pan Jiluan L., Zhang and Renhao H. A new method of controlling the welding arc. *Schweissen und Schneiden*, 10, 1981.

Paton B. E. and Potapevskii A. G. Gas-shielded steady and pulsed-arc welding processes (Review). *Avt. Svarka*, 9, 1973.

Paton B. E. *et al.* Automatic control of pulsed arc welding with a consumable electrode. *Avt. Svarka*, 5, 1965.

Paton, B. E. *et al.* Controlling metal transfer in arc welding with a consumable electrode, *Avt. Svarka*, 5, 1965.

Petrov A. V. Le soudage des tôles minces à l'aide de l'arc pulsé. IIW Doc. XII-375-66, 1-27.

Pfeiffer R. Flammrichten von Aluminium Konstruktionen Dusseldorf Dtsch. *Verlag Schweisstech*, 22: 144-151, 1971.

Pichard C., Poyet P., Guillemot M., Joud J. C. and Desre P. *Compte Rendu du Congrès Physico-chimie et Sidérurgie Versailles*, October 1978. Société Française de Métallurgie, Paris.

Polmear J. The development and commercial evaluation of Al-Zn-Mg alloys containing small additions of silver. *Journal of the Institute of Metals*, 17 (March): 1-15, 1982.

Polmear J. The properties of commercial Al-Zn-Mg alloys. Practical implications of trace additions of silver. *Journal of the Institute of Metals*, 89: 193-202, 1960.

Pomaska H. U. Applications du soudage à l'arc pulsé sous protection gazeuse. *Der Praktiker*, 25 (11): 264-267, 1975.

Popel, Pavlov and Tsarev Skiy. Combined influence of oxygen and sulphur on the surface tension of iron. *Russian Metallurgy*, 4: 42-44, 1975.

Potapevskii V. F. *et al.* Transfer of electrode metal in argo-shielded pulsed arc welding. *Avt. Svarka*, 6, 1965.

Prischi G. Etude préliminaire des nouvelles tendances dans la conception des sources de courant. Doc. XII-F-200-78 de l'IIS.

Puschner P. and Stein H. U. Prozebruckfuhrung zum Verringern der Spritzerbildung beim MAG-Schweiben. *DVS-Berichte Band*, 65.

Puschner P. and Stein H. U. Spritzerfreier Werkstoffubergang beim Schutzgasschweibprozeb durch prozeb-ruckfuhrung. *Industrie Anzeiger*, 101 (32), 1979.

Puschner P. and Stein H. U. Spritzerfreier Werkstoffubergang beim Schutzgasschweibprozeb durch prozeb-ruckfuhrung. *Industrie Anzeiger* 101 (32), 1979.

Rehfeld P. Procédés et matériels d'analyse pour l'études des variations de tension de soudage lors des procédés de soudage électriques, Thèse, TH Hanovre 1969.

Rehfeldt D. and Erdmann-Jesnitzer F. Nouvelles sources de courant du type électronique. Doc. XII-F-197-78 de l'IIS.

Reynolds, M. A. and Harris J. G. Development of a tough, high-strength aluminium alloy with improved stress corrosion resistance. *Aluminium*, 50: 592-596, 1974.

Rinaldi F. and Peloso G. P. Welding vanadium steels for high temperature service. *Brit. Weld. J.*, August: 281-285, 1972.

Rodgers K. J. A study of penetration variability using mechanised TIG welding. *Int. Conf. on The Effects of Residual, Impurity and Micro-Alloying Elements on Weldability and Weld Properties*, London, 15-17 November 1983. The Welding Institute.

Ruckdesche W. E. W. and Smith A. A. Institut International de la Soudure (IIS), Doc. XII-B-170-74 et XII-590-74. *Le Soudage dans le Monde*, 13 (3/4): 65-72, 1975.

Russel A. and Chihoski. The character for stress fields around a weld arc moving on aluminium sheet. *Weld. Int. Suppl.*, 51 (1), 1972.

Salter G. R. and Doherty J. *Metal Construction*, 13 (9): 548-550, 1981.

Salter G. R. and Smith A. A. *British Welding Journal*, 11 (5): 223-228, 1964.

Schellhase M. Source de courant transistorisée à hacheur pour soudage soudage TIG à faible intensité. Doc. XII-F-196-78 de l'IIS.

Schellhase M. and Mannel C. Caractéristiques d'arcs de soudage sous protection de gaz inerte, pour le soudage avec courant pulsé formant ondes carrées. *Schweisstechnik (Berlin)*, 23 (2): 61-64, 1973.

Schellhase M. Module de commande ZIS 835 sur le redresseur de soudage TIG modulé. *Schweisstechnik (Berlin)*, 24 (12): 547-549, 1974.

Schley R., Lacoste J., Kluzinski C. and Piednoir P. Application de Laser Solide Yag en Soudage à l'Instrumentation des Dispositifs d'essai de Sûreté des Elémets Combustibles pour Réacteurs Nucléaires, 3e CISFFEL, September 1983.

Schultz J. P. and Pourrat M. Le soudage TIG multicathode. *Soudage et Techniques Connexes*, 9-10: 341-350, 1976.

Schultz J. P. New possibilities in plasma-arc and dual shielding gas TIG welding. *Welding Fabrication and Surface Treatment. Int. Conf. Singapore*. Singapore Welding Society, July 1980.

Schutz W. and Oberparleiter W. Rissfortschritts and Resfestigkeitseigens-schaften neuer, hodter Aluminium legierungen. *Aluminium*, 48: 734-738, 1977.

Schercliff J. A. Fluid motions due to an electric current source. *J. Fluid Mech.*, 40: 241-250, 1971.

Shimada W. and Gotoh T. Characteristics of high frequency pulsed DC TIG welding process. Institut International de la Soudure, Doc. XII-628-76.

Smith A. A. *Le Soudage dans le Monde*, 8 (1): 28-43, 1970.

Smith A. A. *CO₂ Welding*. Welding Institute, Abington.

Smith A. A. *Australian Welding Journal*, 18 (6): 9-17, 1974.

Smith A. A. *British Welding Journal*, 10 (10): 571-586, 1963.

Smith A. A. Doc. XII-669-77 de l'IIS. *Le Soudage dans le Monde*, 16 (1/2): 25-30, 1978.

Smith E. and Apps K. L. Effect of postweld heat treatment on QT 35 steel. *Brit. Weld. J.*, August: 303-308, 1971.

Sommer A. W., Paton N. E. and Folgner D. G. Effects of thermomechanical treatments on aluminium alloys. *Report AFML-TR-72-5* (AD-748-361), February 1972.

Speidel M. O. Stress corrosion cracking of aluminium alloys. *Metallurgical Transactions A*, 6A (April): 631-651, 1975.

Stalby J. T. Aluminium alloy and process developments for aerospace. *Metals Engineering Quarterly*, 16 (2): 52-57, 1976.

Staley J. T., Hunsicker H. Y. and Schmidt R. New aluminium alloy X7050. TMS Paper Selection, AIME, Paper No. F 71-7.

Staley J. T. and Lyle J. P. Further development of aluminium alloy X7050. Naval Air Systems Command. Contract No. 00019.71.6.0, May 1972.

Taylor A. F. Expériences en soudage TIG avec arc pulsé, faites à UKAEA Springfields. *Research Member Seminar*, London, 7 December 1972 Welding Institute, Abington.

Tesman A. Le préchauffage pour le soudage. *Welding Journal*, August 1982.

Thompson D. S., Subramanya B. S. and Levy S. A. Quench rate effects in Al-Zn-Mg-Cu alloys. *Metallurgical Transactions*, 2 (April): 1149-1160, 1971.

Thompson D. S., Levy S. A. and Spangler G. E. Thermomechanical aging of aluminium alloys. *Aluminium*, 50 (10): 647-649 and (11): 719-722, 1974.

Thompson D. S. Metallurgical factors affecting high strength aluminium alloy production. *Metallurgical Transactions A*, 6A (April): 671-683, 1975.

Thorn K. *et al. Metal Construction*, 14 (3): 128-133.

Ulff C. The effect of stress relieving heat treatments on the mechanical properties of pressure vessel steels. *Jernk. Ann.*, 54: 53-64, 1970.

Van der Torre D. Choix du procédé de soudage. *Colloque sur l'Aluminium et ses Alliages en Construction Soudée*. Institut International de la Soudure, Porto, September 1981.

Viglione J. Fatigue evaluation of the X7050 aluminium alloy. Report NADS-MA-7133, June 1971.

Vinckler A. G. Use of weld simulation tests to study the susceptibility to reheat cracking. *CEGB Welding Conf.*, Southampton, September 1972.

Vinckler A. G. and Pense A. W. A Review of underclad cracking in pressure vessel components. *WRC Bulletin*, 197 (August), 1974.

Watkins B., Vaughan H. G. and Lees G. M. Embrittlement of simulated heat affected zones on low alloy steels. *Brit. Weld. J.*, 13 (6): 350-356.

Watkins B., Wood D. S. and Nichols R. W. Effect of prolonged stress relieving heat treatments on the mechanical properties of reactor pressure vessel steels. *Brit. Weld. J.*, February 1963. Doc. IIS/IIW IX 325-62.

Watkins P. V. C. Welding Institute Research Report 35/10/75, 1975.

Weman K. and Straliv R. Effets de la technologie des semi-conducteurs sur le dééloppement des sources de courant de soudage. Doc. XII-F-193-78 de l'IIS.

Willgoss R. A. and Ali. Laser welding of steels for power plant. *Optics and Laser Technology*, April 1979.

Woods R. A. and Milner D. R. Motion in the weld pool in arc welding. *Weld. J. Res. Suppl.*, 50, 1971.

Zoller: Amélioration des structures soudées en alliages à haute résistance. ETAS CRR 136/77/STI, 97-102.